DISCARDED

DEC 1 0 2024

Asheville-Buncombe
Technical Community College
Learning Resources Center
340 Victoria Road
Asheville, NC 28801

OPTIMISING NEW MODES OF ASSESSMENT:
IN SEARCH OF QUALITIES AND STANDARDS

Innovation and Change in Professional Education

VOLUME 1

Series Editor:
Wim Gijselaers, *Department of Educational Development and Educational Research, University of Maastricht, The Netherlands*

Associate Editors:
LuAnn Wilkerson, *Center of Educational Development and Research, University of California, Los Angeles, CA, U.S.A*
Henny Boshuizen, *Educational Technology Expertise Center, Open University Nederland, The Netherlands*

Editorial Board:
Howard Barrows, *Professor Emeritus, School of Medicine, Southern Illinois University, IL, U.S.A.*
Edwin M. Bridges, *Professor Emeritus, School of Education, Stanford University, CA, U.S.A.*
Thomas M. Duffy, *School of Education, Indiana University, Bloomington, IN, U.S.A.*
Rick Milter, *College of Business, Ohio University, OH, U.S.A.*

SCOPE OF THE SERIES

The primary aim of this book series is to provide a platform for exchanging experiences and knowledge about educational innovation and change in professional education and post-secondary education (engineering, law, medicine, management, health sciences, etc.). The series provides an opportunity to publish reviews, issues of general significance to theory development and research in professional education, and critical analysis of professional practice to the enhancement of educational innovation in the professions.

The series promotes publications that deal with pedagogical issues that arise in the context of innovation and change of professional education. It publishes work from leading practitioners in the field, and cutting edge researchers. Each volume is dedicated to a specific theme in professional education, providing a convenient resource of publications dedicated to further development of professional education.

Optimising New Modes of Assessment: In Search of Qualities and Standards

Edited by

MIEN SEGERS
University of Leiden and University of Maastricht, The Netherlands

FILIP DOCHY
University of Leuven, Belgium

and

EDUARDO CASCALLAR
*American Institutes for Research,
Washington, DC, U.S.A.*

> Asheville-Buncombe
> Technical Community College
> Learning Resources Center
> 340 Victoria Road
> Asheville, NC 28801

KLUWER ACADEMIC PUBLISHERS
DORDRECHT / BOSTON / LONDON

A C.I.P. Catalogue record for this book is available from the Library of Congress.

ISBN 1-4020-1260-8

Published by Kluwer Academic Publishers,
P.O. Box 17, 3300 AA Dordrecht, The Netherlands.

Sold and distributed in North, Central and South America
by Kluwer Academic Publishers,
101 Philip Drive, Norwell, MA 02061, U.S.A.

In all other countries, sold and distributed
by Kluwer Academic Publishers,
P.O. Box 322, 3300 AH Dordrecht, The Netherlands.

Printed on acid-free paper

© 2003 Kluwer Academic Publishers
No part of this work may be reproduced, stored in a retrieval system, or transmitted
in any form or by any means, electronic, mechanical, photocopying, microfilming, recording
or otherwise, without written permission from the Publisher, with the exception
of any material supplied specifically for the purpose of being entered
and executed on a computer system, for exclusive use by the purchaser of the work.

Printed in the Netherlands.

Contents

Contributors	vii
Acknowledgements	ix
Preface	xi
The Era of Assessment Engineering: Changing Perspectives on Teaching and Learning and the Role of New Modes of Assessment MIEN SEGERS, FILIP DOCHY & EDUARDO CASCALLAR	1
New Insights Into Learning and Teaching and Their Implications for Assessment MENUCHA BIRENBAUM	13
Evaluating the Consequential Validity of New Modes of Assessment: The Influence of Assessment on Learning, Including Pre-, Post-, and True Assessment Effects SARAH GIELEN, FILIP DOCHY & SABINE DIERICK	37
Self and Peer Assessment in School and University: Reliability, Validity and Utility KEITH TOPPING	55
A Framework for Project-Based Assessment in Science Education YEHUDIT DORI	89

Evaluating the OverAll Test:
   Looking for Multiple Validity Measures   119
   MIEN SEGERS

Assessment for Learning:
   Reconsidering Portfolios and Research Evidence   141
   ANNE DAVIES & PAUL LEMAHIEU

Students' Perceptions about New Modes of Assessment in Higher
   Education: a Review   171
   KATRIEN STRUYVEN, FILIP DOCHY & STEVEN JANSSENS

Assessment of Students' Feelings of Autonomy, Competence, and
   Social Relatedness:
   A New Approach to Measuring the Quality of the Learning
   Process through Self- and Peer Assessment   225
   MONIQUE BOEKAERTS & ALEXANDER MINNAERT

Setting Standards in the Assessment of Complex Performances:
   The Optimized Extended-Response Standard Setting Method   247
   ALICIA CASCALLAR & EDUARDO CASCALLAR

Assessment and Technology   267
   HENRY BRAUN

Index   289

Contributors

Menucha Birenbaum, *School of Education, Tel Aviv University, Ramat Aviv 69978, Israel.* *biren@post.tau.ac.il*

Monique Boekaerts, *Leiden University, Center for the Study of Education and Instruction, Postbus 9500, 2300 RA Leiden, The Netherlands.* *boekaert@fsw.LeidenUniv.nl*

Henry Braun, *Educational testing Service, Rosedale Road, Princeton, NJ 08541, USA.* *hbraun@ets.org*

Alicia Cascallar, *Assessment Group International, 6030 Kelsey Court, Falls Church, VA 22044, USA.*

Eduardo Cascallar, *American Institutes for Research, 1000 Thomas Jefferson Street, N.W., Washington, DC 0007, USA.* *ECascallar@air.org*

Anne Davies, *Classroom Connections International, 2449D Rosewall Crescent Courtenay, B.C., V9N 8R9, Canada.* *anne@connect2learning.com*

Sabine Dierick, *University Maastricht, Faculty of Law, Department of Educational Innovation and Information Technology, PO Box 616, 6200 MD Maastricht, The Netherlands.*

Filip Dochy, *University of Leuven, Department of Instructional Science, Centre for Research on Teacher and Higher Education, Vesaliussstraat 2, 3000 Leuven, Belgium.* *filip.dochy@ped.kuleuven.ac.be*

Yehudit J. Dori, *Department of Education in Technology and ScienceTechnion, Israel institute of Technology, Haifa 32000, Israel*
And
Center for Educational Computing Initiatives, Massachusetts Institute of Technology, Cambridge, MA 02139-4307, USA. dori@ceci.mit.edu

Sarah Gielen, *University of Leuven, Department of Instructional Science, Centre for Research on Teacher and Higher Education, Vesaliussstraat 2, 3000 Leuven, Belgium.* sarah.gielen@ped.kuleuven.ac.be

Steven Janssens, *University of Leuven, Department of Instructional Science, Centre for Research on Teacher and Higher Education, Vesaliussstraat 2, 3000 Leuven, Belgium.*

Paul LeMahieu, *Senior Research Associate, National Writing Project, University of California, Berkeley, USA.*

Alexander Minnaert, *Leiden University, Center for the Study of Education and Instruction, Postbus 9500, 2300 RA Leiden, The Netherlands.*

Mien Segers, *University Maastricht, Department of Educational Development and Research, PO Box 616, 6200 MD Maastricht, The Netherlands.* m.segers@educ.unimaas.nl

Katrien Struyven, *University of Leuven, Department of Instructional Science, Centre for Research on Teacher and Higher Education, Vesaliussstraat 2, 3000 Leuven, Belgium.*
katrien.struyven@ped.kuleuven.ac.be

Keith Topping, *Department of Psychology, University of Dundee, Dundee, DD1 4HN Scotland.* k.j.topping@dundee.ac.uk

Acknowledgements

Working with so many experts in the field of new modes of assessment provided a challenging experience. We are grateful that they responded enthusiastically to our request and contributed a chapter. They all contributed greatly to the successful completion of this book.

We would like to give special thanks to Prof. Kari Smith (Oranim Academic College of Education, Israel) and Prof. J. Ridgway (University of Durham, UK) for their constructive reviews. Their positive comments as well as their suggestions for improvement were helpful in finishing our work.

We are grateful to the editorial board of the book series 'Innovation and Change in Professional Education" for giving us the opportunity to bring together the research of experts in the field of new modes of assessment.

Henny Dankers deserves special recognition for her diligent work on the layout of this book and for maintaining a positive attitude, even when messages from all over the world forced her to change schedules. Also a word of thanks for Bob Janssen Steenberg, our student-assistant, who was always willing to give a helping hand when deadlines came close.

Mien Segers
Filip Dochy
Eduardo Cascallar

Maastricht
January 2003

Preface

French novelist Marcel Proust instructs us that, "a voyage of discovery consists, not of seeking new landscapes, but of seeing through new eyes." Nowhere in the practice of education do we need to see through new eyes than in the domain of assessment. We have been trapped by our collective experiences to see a limited array of things to be assessed, a very few ways of assessing them, limited strategies for communicating results and inflexible roles of players in the assessment drama.

This edited book of readings jolts us out of traditional habits of mind about assessment. An international team of innovative thinkers relies on the best current research on learning and cognition, to describe how to use assessment to promote, not merely check for, student learning. In effect, they explore a new vision of assessment for the new millennium. The authors address the rapidly expanding array of achievement targets students must hit, the increasingly productive variety of assessment methods available to educators, innovative ways of collecting and communicating evidence of learning, and a fundamental redefinition of both students' and teachers' roles in the assessment process.

With respect to the latter, special attention is given throughout to what I believe to be the future of assessment in education: assessment FOR learning. The focus is on student-involvement in the assessment, record keeping and communication process. The authors address not only critically important matters of assessment quality but also issues related to the impact of assessment procedures and scores on learners and their well being.

Those interested in the exploration of assessments that are placed in authentic performance contexts, multidimensional in their focus, integrated into the learning process and open to the benefits of student involvement will learn much here.

Rick Stiggins
Assessment Training Institute
Portland, Oregon USA
January 10, 2003

The Era of Assessment Engineering: Changing Perspectives on Teaching and Learning and the Role of New Modes of Assessment

Mien Segers[1], Filip Dochy[2] & Eduardo Cascallar[3]
[1]*Department of Educational Development and Research, University Maastricht, The Netherlands,*
[2]*University of Leuven, Department of Instructional Science, Centre for Research on Teacher and Higher Education, Belgium,* [3]*American Institutes for Research, USA*

1. INTRODUCTION

Assessment of student achievement is changing, largely because today's students face a world that will demand new knowledge and abilities, and the need to become life-long learners in a world that will demand competencies and skills not yet defined. In this 21st century, students need to understand the basic knowledge of their domain of study, but also need to be able to think critically, to analyse, to synthesize and to make inferences. Helping students to develop these skills will require changes in the assessment culture and the assessment practice at the school and classroom level, as well as in higher education, and in the work environment. It will also require new approaches to large-scale, high-stakes assessments.

A growing body of literature describes these changes in assessment practice and the development of new modes of assessment. However, only a few of them address the critical issue of quality. The paradigm change from a testing culture to an assessment culture can only be continued when research offers sufficient empirical evidence for the various aspects of quality of this new assessment culture (Birenbaum & Dochy, 1996). This book intends to contribute to this aim by presenting a series of studies on various aspects of quality of new modes of assessment. It starts by elaborating on the conceptual framework of this paradigm change and related new approaches in edumetrics related to the development of an expanded set of quality criteria. A series of new modes of assessment with special emphasis on quality considerations involved in them, are described from a research perspective. Finally, recent developments in the field of e-assessment and the impact of new technologies will be described. The current chapter will introduce the different issues addressed in this book.

2. CHANGING PERSPECTIVES ON THE NATURE OF ASSESSMENT WITHIN THE LARGER FRAMEWORK OF LEARNING THEORIES

During the last decades, the concept of learning has been reformulated based on new insights, developed within various related disciplines such as cognitive psychology, learning sciences and instructional psychology. Effective or meaningful learning is conceived as occurring when a learner constructs his or her own knowledge base that can be used as a tool to interpret the world and to solve complex problems. This implies that learners must be self-dependent and self-regulating, and that they need to be motivated to continually use and broaden their knowledge base. Learners need to develop strategic learning behaviour, meaning they must master effective strategies for their own learning. Finally, learners need meta-cognitive skills in order to be able to reflect on their own and others' perspectives.

These changes in the current views on learning lead to the rethinking of the nature of assessment. Indeed, there is currently a large agreement within the field of educational psychology as well as across its boundaries that learning should be in congruence with assessment (Birenbaum & Dochy, 1996). This has lead to the raise of the so-called assessment culture. The major changes in assessment, as defined by Kulieke, Bakker, Collins, Fennimore, Fine, Herman, Jones, Raack, & Tinzmann (1990) are moving from testing to multiple assessments, and from isolated to integrated assessment.

We can portray the aspects of assessment in seven continua. This schema (figure 1) is mainly based on Kulieke et al. (1990, p.5).

The Era of Assessment Engineering

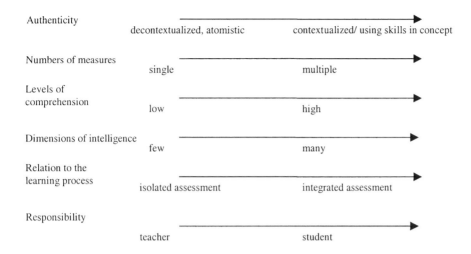

Figure 1. The Characteristics of Assessment on Six Continua

The first continuum shows a change from decontextualized, atomic tests to authentic contextualized tests. In practice, it refers to the shift from the so-called objective tests with item formats such as short answer, fill-in blank, multiple-choice and true/false to the use of portfolio assessment, project-based assessment, performance assessment, etc. The second continuum shows a tendency from describing a student's competence with one single measure (a mark) towards portraying a student's competence by a student's profile based on multiple measures. The third continuum depicts the movement from low levels of competence towards high levels of competence. This is the move from mainly assessing reproduction of knowledge to assessing higher-order skills. The fourth continuum refers to the multidimensionality of intelligence. Intelligence is more than cognition; it implies certainly meta-cognition, but also affective and social dimensions and sometimes psychomotor skills. The fifth continuum concerns the move towards integrating assessment into the learning process. To a growing extent, the strength of assessment as a tool for dynamic ongoing learning is stressed. The sixth continuum refers to the change in responsibilities, not only in the learning process but also in the assessment process. The increasing implementation of self- and peer assessment are examples of this move from teacher to student responsibility.

Finally, the seventh continuum refers to the shift from the assessment of learning towards an equilibrated assessment of learning and assessment for learning. Research has shown convincingly that using assessment as a tool for learning, including good and well-timed feedback, leads to better results when assessing learning outcomes.

The chapter of Birenbaum elaborates on this paradigm change in learning and assessment. She describes the current perspectives on instruction, learning and assessment (ILA) and illustrates this ILA culture with an example of a learning environment.

3. EDUMETRICS AND NEW MODES OF ASSESSMENT

With the increasing implementation of new modes of assessment at all levels of education, from state and district assessments to classroom assessments, questions are raised about the quality of these new modes of assessment (Birenbaum & Dochy, 1996). Edumetric indicators like reliability and validity are used traditionally to evaluate the quality of educational assessment. The validity question refers to the extent to which assessment measures what it purports to measure. Does the content of assessment correlate with the goals of education? Reliability was traditionally defined as the extent to which a test measures consistently. Consistency in test results demonstrated objectivity in scoring: the same results were obtained if the test was judged by another person or by the same person at another time. The meaning of the concept reliability was determined by the then prevailing opinion that assessment needs to fulfil above all a selecting function. Fairness in testing was aligned with objectivity. Striving to achieve objectivity in testing and comparing scores resulted in the use of standardized testing forms, like multiple-choice tests. Some well-known scientists state these days that since we uncritically searched for the most reliable tests, learning processes of children and students in schools are not what we hoped for, or what they need to be. Tests have an enormous power in steering learning processes. This might work to such an extent that even very reliable tests do elicit unproductive or unwanted learning. We can think here of students trying to memorize old test items and their answers, students anticipating expected test formats, students getting drilled and practicing in guessing, etc. At the EARLI Assessment conference in the UK (Newcastle, 2002) a colleague said: "Psychometrics is not God; valid and reliable tests do not warrant good learning processes, they do not guarantee that we get where we want to. Even worse, the most perfect tests could lead to the worst (perhaps superficial) learning".

Various researchers like Messick, Linn, Baker and Dunbar have criticized the inappropriateness of the traditional psychometric criteria for evaluating the quality of new modes of assessment. They pointed out that there was an urgent need for the development of a different or an expanded set of quality criteria for new modes of assessment. Although there are differences in perspectives between the researchers, four aspects are considered as part of a comprehensive strategy for conducting a quality evaluation: the validity of the assessment tasks, the validity of assessment scoring, the generalizability of the assessment, and the consequential validity of the assessment process. Because of research evidence regarding the steering effect of assessment, and indicating the power of assessment as a tool for learning, the consequential aspect of validity has gained increasing interest. The central question is to what extent does an assessment lead to the intended consequences or does it produce unintended consequences such as test anxiety and teaching to the test. The chapter of Gielen, Dochy & Dierick elaborates on these quality issues. They illustrate the consequential validity of new modes of assessment from a research perspective. The effect of various aspects of new modes of assessment on student learning are described: the effect of the cognitive complexity of assessment tasks, of feedback (formative function of assessment or assessment for learning), of transparent assessment criteria and the involvement of students in the assessment process, and the effect of criterion-referenced standards setting.

4. THE QUALITIES OF NEW MODES OF ASSESSMENT

Since there is a significant consensus about the main features of effective learning, and the influence of assessment on student learning, on instruction and on curriculum is widely acknowledged, educators, policy makers and others are turning to new modes of assessment as part of a broader educational reform. The movement away from traditional, multiple-choice tests to new modes of assessment has included a wide variety of instruments.

In alignment with the principle that a student should be responsible for his own learning and assessment process, self-assessment strategies are implemented. Starting from the perspective that learning is a social process and self-reflection is enriched by critical reflection by peers, peer-assessment is now widely used. Stressing the importance of meta-cognitive skills, student responsibility and the complex nature of competencies, portfolios are implemented in a variety of disciplines and at different school levels. One of the widely implemented reforms in the instructional process is project-based education. In alignment with this instructional approach and in order to integrate instruction, learning and assessment (ILA), many schools developed project-based assessments. Finally, the emphasis on problem solving and on the use of a variety of authentic problems in order to stimulate transfer of knowledge and skills has lead to the development of case-based assessment instruments such as the OverAll Test.

The chapters of Topping, Boekaerts and Minnaert, Davies & Le Mahieu, Dori and Segers will elaborate on these new modes of assessment. They will present and discuss research studies investigating different quality aspects.

Topping addresses the issues of validity in scoring and consequential validity of self- and peer assessment. He concludes that the validity in scoring of self-assessment is lower in comparison with professional teachers and more variable. There is more substantial hard evidence that peer assessment can result in improvements in the effectiveness and quality of learning, which is at least as good as gains from teacher assessment, especially in relation to writing. In other areas, the evidence is softer. Of course, self and peer assessment are not dichotomous alternatives - one can lead to and inform the other. Both can offer valuable triangulation in the assessment process and both can have measurable formative effects on learning, given good quality implementation. Both need training and practice, arguably on neutral products or performances, before full implementation, which should feature monitoring and moderation. The chapter of Boekaerts and Minnaert adds an alternative perspective to the study of the qualities of new modes of assessment. Research (Boekaerts, 2002) indicates that motivation factors are powerfully present in any form of assessment and bias the students' judgment of their own or somebody else's performance. In a recent study on the impact of affect on self-assessment, Boekaerts (2002, in press) showed that students' appraisal of the demand capacity ratio of a mathematics task, before starting on the task, contributed a large proportion of the variance explained in their self-assessment at task completion. Interestingly, the students' affect (experienced positive and negative emotions during the math task) mediated this effect. Students who experienced intense negative emotions during the task underrated their performance while students who experienced positive emotions, even in addition to negative emotions, overrated their performance. This finding is in line with much research in mainstream psychology that has demonstrated the effect of positive and negative mood state on performance. In light of these results, it is surprising that literature on assessment and on qualities of new modes of assessment mainly focus on the assessment of students' performances on product as well as on process level. They do not take into account the assessment of potential intervening factors such as students' interest. The chapter of Boekaerts and Minnaert offer insight in the relation between three basic psychological needs and interest during the different phases of a cooperative learning process. For educators, the research results are informative for enhancing the accuracy of the diagnosis of students' performances and their self-assessment. Additionally, the authors present an interesting method for self-assessment of students' interest and the underlying psychological needs. For researchers in the field of assessment, the research results presented indicate the importance of measuring student's interest and its underlying psychological needs in order to interpret in a more accurate way the validity of students' self-assessments.

Based on a series of research studies conducted in science education, Dori presents a framework for project-based assessment. The projects are interpreted as an ongoing process integrating instruction, learning and assessment. The project-based assessment is considered as suited for fostering and evaluating higher-order thinking skills. In the three studies presented, the assessment comprises several assessment instruments, including portfolio assessment, community expert assessment by observation, self-assessment, and a knowledge test. The findings of the three studies indicate that project-based assessment indeed fosters higher-order thinking skills in comparison with students who experienced traditional learning environments.

Davies & le Mahieu present examples of studies about various quality issues of portfolio assessment. Studies indicate that portfolios impact positively on learning in terms of increased student motivation, ownership, and responsibility. Researchers studying portfolios found that when students choose work samples, the result is a deeper understanding of content, a clearer focus and better understanding of quality product. Portfolio construction involves skills such as awareness of audience, awareness of personal learning needs, understanding of criteria of quality and the manner in which quality is revealed in their work and compilations of it as well as development of skills necessary to complete a task. Besides the effect of portfolio-assessment on learning as an aspect of the consequential validity, Davies and le Mahieu explore the effect on instruction in a broader context. There is evidence that portfolios enrich conversation about learning and teaching with students as well as the parents involved. Portfolios designed to support schools, districts, and cross-district learning (such as provincial or state-level assessment) reflect more fully the kinds of learning being asked of students in today's schools and support teachers and other educators in learning more about what learning can look like over time. Looking at empirical evidence for the validity in scoring, inter-rater reliability of portfolio work samples continues to be a concern. The evaluation and classification of results is not simply a matter of right and wrong answers, but of inter-rater reliability, of levels of skill and ability in a myriad of areas as evidenced by text quality and scored by different people, a difficult task at best. Clear criteria and anchor papers assist the process. Experience seems to improve inter-rater reliability.

The OverAll Test can be seen as an example of case-based assessment, widely used in curricula where problem solving is one of the core goals of learning and instruction. The research study presented in the chapter of Segers explores various validity aspects of the OverAll test. Although there is evidence for alignment of the OverAll Test with the curriculum and results findings indicate criterion validity, the results of the study of the consequential validity indicate some concern. A survey and semi-structured interviews with students and staff indicate intended as well as unintended effects of the OverAll test on the way students learn, on the perceptions of students and teachers of the goals of education and, in particular, assessment. From these results, it is clear that, more than the objective learning environment, the subjective learning environment, as perceived by the students, plays an important role in the effect of the OverAll Test on students' learning. There is evidence that in most learning environments, the subjective learning environment plays a major role in determining the influence of pre-, post en true assessment effects in learning. Hence, there is a lot of work in what we should call "assessment engineering".

5. STUDENTS' PERCEPTIONS OF NEW MODES OF ASSESSMENT

As it is indicated in the chapter of Segers, student learning is subject to a dynamic and richly complex array of influences which are both direct and indirect, intentional and unintended (Hounsell, 1997). Entwistle (1991) found that the factual curriculum, including assessment demands, does not direct student learning, but the students' perceptions. This means that investigating the reality as experienced by the students can be a value-added to gain insight into the effect of assessment on learning.

It is widely acknowledged that new modes of assessment can contribute to effective learning (Birenbaum & Dochy, 1996). In order to gain insight into the underlying mechanism, it seems to be worthwhile to investigate students' perceptions.

This leads to the question: how do students perceive new modes of assessment and what are the influences on their learning. The review study of Struyven, Dochy & Janssen as presented in this book evidenced that students' perceptions of assessment and its properties have considerable influences on students' approaches to learning and, more generally, on student learning. Vice versa, students' approaches to learning influence the way in which students perceive assessment. Research studies on the relation between perceptions of assessment and student learning report on a variety of assessment formats such as self- and peer-assessment, portfolio assessment and OverAll Assessment. Aspects of learning taken into consideration are for example the perceived level of test anxiety, the perceived effect on self-reflection and on engagement in the learning process, the perceived effect on structuring the learning process, and the perceived effect on deep-level learning. The integration of the assessment in the learning and instruction process seems to play a mediating role in the relation between perceptions of assessment and effects on learning. Furthermore, it was found that students hold strong views about different formats and methods of assessment. For example, within conventional assessment, multiple choice format exams are seen as favourable assessment methods. But when conventional assessment and alternative assessment methods are discussed and compared, students perceive alternative assessment as being more "fair" than traditional assessment methods. From students' point of view, assessment has a positive effect on their learning and is fair when it (Sambell, McDowell, & Brown, 1997):

- Relates to authentic tasks.
- Represents reasonable demands.
- Encourages students to apply knowledge to realistic contexts.
- Emphasizes the need to develop a range of skills.
- Is perceived to have long-term benefits.

6. SETTING STANDARDS IN CRITERION-REFERENCED PERFORMANCE ASSESSMENTS

Many of the new modes of assessment, including so-called "authentic assessments", address complex behaviours and performances that go beyond the usual multiple-choice tests. This is not to say that objective testing methods cannot be used for the assessment of these complex abilities and skills, but constructed response methods many times present a practical alternative. Setting of defensible, valid standards becomes even more relevant for the family of constructed response assessments, which include extended-response instruments.

Several methods to carry out standard settings on extended-response examinations have been used. In order to also deal with multidimensional scales that can be found in extended response examinations the Optimized Extended-Response Standard Setting method (OER) was developed (Schmitt, 1999). The OER standard setting method uses well-defined rating scales to determine the different scoring points where judges will estimate minimum passing points for each scale. Recent conceptualisations, such as those differentiating between criterion- and construct-referenced assessments (William, 1997), present very interesting distinctions between the descriptions of levels and the domains. The method described in the chapter by Cascallar & Cascallar can integrate the conceptualisation, as providing both an adequate "description" of the levels, as attained by the "informed consensus" of the judges (Schmitt, 1999), as well as a flexible "exemplification" of the level inherent in the process to reach the consensus. As it has been pointed out, there is an essential need to estimate the procedural validity of judgment-based cut-off scores. The OER Standard Setting Method suggests a methodology and provides the procedures to maintain the necessary degree of consistency to make critical decisions that affect examinees in the different settings in which their performance is measured against the cut scores set using standard setting procedures. With reliability being a necessary but not sufficient condition for validity, it is necessary to investigate and establish valid methods for the setting of those cut-off points (Plake & Impara, 1996). The general uneasiness with the current standard setting methods (Pellegrino, Jones, & Mitchell, 1999) rests largely on the fact that setting standards is a judgment process that needs well-defined procedures, well-prepared judges, and the corresponding validity evidence. This validity evidence is essential to reach the quality commensurate with the importance of its application in many settings (Hambleton, 2001). Ultimately, the setting of standards is a question of values and of the decision-making involved in the evaluation of the relative weight of the two types of errors of classification. This chapter addresses these issues and describes a methodology to attain the necessary level of quality in this type of criterion-referenced assessments.

7. THE FUTURE OF NEW MODES OF ASSESSMENT IN AN ICT-WORLD

It is likely that the widespread implementation of ICT in the world of education will leave its marks on instruction, learning and assessment. Braun, in his chapter, presents a framework for the analysis of forces like technology, shaping the practice of assessment, comprising three dimensions: context, purpose and assets. The direct effect of technology mainly concerns the assets dimension, influencing the whole process of assessment design and implementation. Technology increases the efficiency and effectiveness of identifying the set of feasible assessment designs, constructing and generating assessment tasks, delivering authentic assessment tasks in a flexible way with various levels of interactivity and automated scoring of students constructed responses to the assessment tasks. In this respect, technology can enhance the validity of new modes of assessment, also in large-scale assessment contexts. Indirectly, technology has an effect on the development of many disciplines, for example cognitive science, gradually influencing the constructs and models that influence the design of assessments for learning. Additionally, there is a complex interplay between technology on the one hand and the political, economical and market forces on the other hand.

8. ASSESSMENT ENGINEERING

It is obvious that the science of assessment in its current meaning, referring to new modes of assessment, assessment for learning, assessment of competence, is still in an early phase. Certainly, there is a long way to go, but research results point in the same conclusive direction: the effects of assessment modalities, assessment formats, and the influence of subjective factors in assessment environments are not to be underestimated. Surely, recent scientists have given clear arguments for further developments in this direction and for avoiding earlier pitfalls such as concluding that assessments within learning environments are largely comparable with assessment of human intelligence and other psychological phenomena. Also within the area of edumetrics, a lot of research is needed in order to establish a well-defined but evolving framework, and the corresponding instruments within a sound quality assurance policy. The science of assessment engineering, trying to fill the gaps we find in aligning learning and assessment, requires more research within many different fields. The editors of this book hope that this contribution will be another step forward in the field. Nexon, my horse, come here; I will take another bottle of Chateau La Croix, and we will ride further in the wind.

REFERENCES

Birenbaum, M., & Dochy, F. (Eds.). (1996). *Alternatives in Assessment of Achievement, Learning Processes and prior Knowledge.* Boston: Kluwer Academic.

Boekaerts, M. (2002, in press). Toward a Model that integrates Affect and Learning, Monograph published by *The British Journal of Educational Psychology.*

Entwistle, N. J. (1991). Approaches to learning and perceptions of the learning environment. Introduction to the special issue. *Higher Education, 22,* 201-204.

Hambleton, R. K. (2001). Setting performance standards on educational assessments and criteria for evaluating the process. In G.J. Cizek (Ed.), *Setting performance standards: Concepts, methods, and perspectives.* Mahwah, NJ: Lawrence Erlbaum Publishers

Hounsell, D. (1997a). Contrasting conceptions of essay- writing. In F. Marton, D. Hounsell, & N. Entwistle (Eds.), *The experience of learning. Implications for teaching and studying in higher education* [second edition] (pp. 106-126). Edinburgh: Scottish Academic Press.

Hounsell, D. (1997b). Understanding teaching and teaching for understanding. In F. Marton, D. Hounsell, & N. Entwistle (Eds.), *The experience of learning. Implications for teaching and studying in higher education* [second edition] (pp. 238-258). Edinburgh: Scottish Academic Press.

Kulieke, M., Bakker, J., Collins, C., Fennimore, T., Fine, C., Herman, J., Jones, B.F., Raack, L., & Tinzmann, M. B. (1990). *Why Should Assessment Be Based on a Vision of Learning?* North Central Regional Educational Laboratory (NCREL), Oak Brook.

Linn, R. L., Baker, E., & Dunbar, S. B. (1991). Complex, performance-based assessment: Expectations and validation criteria. *Educational Researcher, 16*, 1-21.

Messick, S. (1995). Standards of Validity and the Validity of Standards in Performance Assessment. *Educational Measurement: Issues and Practices, Winter 1995*, 5-8.

Plake, B. S., & Impara, J. C. (1996). *Intrajudge consistency using the Angoff standard setting method.* Paper presented at the annual meeting of the National Council on Measurement in Education. New York, NY.

Pellegrino, J. W., Jones, L. R., & Mitchell, K. J. (Eds.). (1999). *Grading the nation's report card.* Washington, DC: National Academy Press.

Sambell, K., McDowell, L., & Brown, S. (1997). "But is it fair?": an exploratory study of student perceptions of the consequential validity of assessment. *Studies in Educational Evaluation, 23* (4), 349-371.

Schmitt, A. (1999). *The Optimized Extended Response Standard Setting Method.* Technical Report, Psychometric Division. Albany, NY: Regents College.

William, D. (1997). *Construct-referenced assessment of authentic tasks: alternatives to norms and criteria.* Paper presented at the 7th Conference of the European Association for Research in Learning and Instruction. Athens, Greece. August 26-30.

New Insights Into Learning and Teaching and Their Implications for Assessment

Menucha Birenbaum
Tel Aviv University, Israel

Key words: This chapter is based on a keynote address given at the First Conference of the Special Interest Group on Assessment and Evaluation of the European Association for Learning and Instruction (EARLI), Maastricht, The Netherlands, Sept. 13, 2000.

1.   INTRODUCTION

Instruction, learning and assessment are inextricably related and their alignment has always been crucial for achieving the goals of education (Biggs, 1999). This chapter examines the relationships among the three components –instruction, learning, and assessment - in view of the challenge that education is facing at the dawn of the 21st century. It also attempts to identify major intersections where research is needed to better understand the potential of assessment for improving learning.

The chapter comprises three parts: the first examines the assumptions underlying former and current perspectives on instruction, learning, and assessment (ILA). The second part describes a learning environment – an ILA culture - that illustrates the current perspective. The last part suggests new directions for research on assessment that is embedded in such a culture.

Although the chapter is confined to instruction, learning, and assessment in higher education, much of what is discussed is relevant and applicable to all levels of education. The term assessment (without indication of a specific type) is used throughout the chapter to denote formative assessment, also known as classroom assessment or assessment for learning.

2. THEORETICAL FRAMEWORK

2.1 The Challenge

Characteristic of life at the dawn of the 21st century are rapid changes - political, economical, social, esthetical and ethical - as well as rapid developments in science and technology. Communication has become the infrastructure in the post-industrial society and advances in this area, especially in ICT (information communication technology), have changed the scale of human activities. Moreover, new conceptions of time and space have transcended geographical boundaries and thereby accelerating the globalisation process (Bell, 1999). This era, known too as the "knowledge age", is also characterized by the rapidly increasing amount of human knowledge, which is expected to go on growing at an even faster pace. Likewise, due to the advances in ICT the volume of easily accessed information is rapidly increasing. Consequently, making quick adjustments and being a life long learner (LLL) are becoming essential capabilities, now more than ever, for effective functioning in various areas of life (Glatthorn & Jailall, 2000; Jarvis, Holford, & Griffin, 1998; Pintrich, 2000).

In order to become a life long learner one has to be able to regulate one's learning. There are many definitions of self-regulated learning but it is commonly agreed that the notion refers to the degree that students are metacognitively, motivationally and behaviourally active in their learning (Zimmerman, 1989). The cognitive, metacognitive and resource management strategies that self-regulated learners activate in combination with related motivational beliefs help them accomplish their academic goals and overcome obstacles along the way (Pintrich, 2000; Randi & Corno, 2000; Schunk & Zimmerman, 1994). The need to continue learning throughout life, together with the increasing availability of technological means for participating in complex networks of information, resources, and instruction, highly benefit self-regulated learners. They can assume more responsibility for their learning by deciding what they need to learn and how they would like to learn it.

Bell (1999) notes that "the post industrial society deals with fundamental changes in the *techno-economic* sphere and has its greatest impact in the areas of education and work and occupations that are the centres of this sphere." (p. lxxxiii). Indeed a brief look at the employment ads in the weekend newspapers is enough to give an idea of the rapid changes that are taking place in the professional working place. These changes mark the challenge higher education institutes face, having to prepare their students to become professional experts in the new working place. Such experts are required to create, apply and disseminate knowledge and continuously construct and reconstruct their expertise in a process of life-long learning. They are also required to work in teams and to cooperate with experts in various fields (Atkins, 1995; Tynjälä, 1999).

However up-to-date, many higher education institutes do not seem to be meeting this challenge. There is a great deal of evidence indicating that many university graduates acquire only surface declarative knowledge of their discipline rather than deep conceptual understanding so that they lack the capacity of thinking like experts in their areas of study (Ramsden, 1987). Furthermore, it has been noted that traditional degree examinations do not test for deep conceptual understanding (Entwistle & Entwistle, 1991).

But having specified the challenge, the question is how it can be met. The next part of this chapter reviews current perspectives of learning, teaching and assessment that offer theoretical foundations for creating powerful learning environments that afford opportunities for promoting the required expertise.

2.2 Paradigm Change

It is commonly acknowledged that in order to meet the goals of education an alignment or a high consistency between instruction, learning and assessment (ILA) is required (Biggs, 1999). Such an alignment was achieved in the industrial era. The primary goal of public education at that time was to prepare members of social classes that had previously been deprived of formal education for efficient functioning as skilled workers at the assembly line. Public education therefore stressed the acquisition of basic skills while higher order thinking and intellectual pursuits were reserved for the elite ruling class. The ILA practice of public education in the industrial era can be summarized as follows: instruction-wise: knowledge transmission; learning-wise: rote memorization; and assessment-wise: standardized testing. The teacher in the public school was perceived as the "sage on the stage" who treated the students as "empty vessels" to be filled with knowledge. Freire (1972) introduced the "banking" metaphor to describe this educational approach, where the students are depositories and the teacher - the depositor. Learning that fits with this kind of instruction is carried out through tedious drill and practice, rehearsals and repetitions of what was taught in class or in the textbook. The aligned assessment approach is a quantitative one aimed at differentiating among students and ranking them according to their achievement. This is done by utilizing standardized tests comprising decontextualized, psychometrically designed items of the choice-response format that have a single correct answer and test mainly low-level cognitive skills. As to the responsibilities of the parties involved in the assessment process, the examinees neither participate in the development of the test items nor in the scoring process, which remains a mystery to them. Moreover, under this testing culture, instruction and assessment are considered separate activities, the former being the responsibility of the teacher and the latter the responsibility of the measurement expert.

Theoreticians focusing on *theories of mind* argue that educational practices are premised on a set of beliefs about learners' minds, which they term "folk psychology" (Olson & Bruner, 1997). They term the processes required to advance knowledge and understanding in learners "folk pedagogy". Olson and Bruner (1997) claim that teachers' folk pedagogy reflects their folk psychology. They distinguish four models of learners' minds and link them to models of learning and teaching. One of these models conceptualises the learner as a knower. The folk psychology in this case conceives the learner's mind as a *tabula rasa* equipped with the ability to learn. The mind is conceived as passive (i.e., a vessel waiting to be filled) and any knowledge deposited into it is seen as cumulative. The corresponding folk pedagogy conceives the instructional process as managing the learner from the outside (i.e., performing teaching by telling.) The resemblance between this conceptualisation and the traditional perspective on ILA described above is quite obvious. Olson and Bruner classify this model as an externalist theory of mind, meaning that its focus is on what the teacher can do to foster learning rather than on what the learners can do or intend to do. Also implied is a disregard on the part of the teacher for the way the learners see themselves, thus aspiring, so Olson and Bruner claim, to the objective, detached view of the scientist.

Indeed, the traditional perspective on ILA is rooted in the empirical-analytical paradigm that dominated western thinking from the mid-18th to the mid-20th century. It reflects an empiricist (positivist) epistemological stance according to which knowledge is located outside the subject (i.e., independent of the knower) and only one reality/truth exists. It is objectively observable through the senses and therefore it must be discovered rather than created (Cunningham & Fitzgerald, 1996; Guba, 1990). The traditional perspective on ILA is also in line with theories of intelligence and learning that share the empirical-analytic paradigm. These theories stress the innate nature of intelligence and measure it as a fixed entity that is normally distributed in the population. The corresponding theories of learning are the behaviourist and associationist (connectionist) ones.

As was mentioned above, the goals of education in the knowledge age have changed, therefore requiring a new perspective for ILA. Indeed such a perspective is already emerging. It is rooted in the interpretative or constructivist paradigm and reflects the poststructuralist and postmodernist epistemologies that have dominated much of the discourse on knowledge in the western world in the past three decades (Cunningham & Fitzgerald, 1996). According to these epistemologies, there are as many realities as there are knowers. If truth is possible it is relative (i.e., true for a particular culture). Knowledge is social and cultural and does not exist outside the individuals and communities who know it. Consequently, all knowledge is considered to be created/constructed rather than discovered.

The new perspective on ILA is also rooted in new theories of human intelligence that stress the multidimensional nature of this construct (Gardner, 1983, 1993; Sternberg, 1985) and the fact that it should not be treated as a fixed entity. There is evidence that training interventions can substantially raise the individual's level of intellectual functioning (Feuerstein, 1980; Sternberg, 1986). According to these theories, intelligence is seen as mental self-management (Sternberg, 1986) implying that one can learn how to learn. Furthermore, mental processes are believed to be dependent upon the social and cultural context in which they occur and to be shaped as the learner interacts with the environment.

What then is the nature of the emerging perspective on ILA? In order to answer this question we first briefly review the assumptions underlying current perspectives on learning and the principles derived from them.

2.2.1 Current Perspectives on Learning

Constructivism is the umbrella under which learning perspectives that focus on mind-world relations are commonly grouped. These include modern (individual) and post-modern (social) learning theories (Prawat, 1996). The former, also referred to as cognitive approaches, focus on the structures of knowledge in learners' minds and include, among others, the cognitive-schema theory (Derry, 1996), Piaget-based radical constructivism (Von Glasersfeld, 1995) and the constructivist revision of information processing theory (Mayer, 1996). Post-modern social constructivist theories on the other hand, reject the notion that the locus of knowledge is in the individual. The approaches referred to as situative emphasize the distributed nature of cognition and focus on students' participation in socially organized learning activities (Brown, Collins, & Duguid, 1989; Lave & Wenger, 1991). Social constructivist theories include, among others, socio-cultural constructivism in the Vygotzkian tradition (Vygotsky, 1978), symbolic interactionism (Blumer, 1969; Cobb & Yackel, 1996), Deweyan idea-based social constructivism (Dewey, 1925/1981) and the social psychological constructivist approach (Gergen, 1994).

Common to all these perspectives is the central notion of activity - the understanding that knowledge, whether public or individual, is constructed. Yet they vary with respect to their assumptions about the nature of knowledge (Phillips, 1995) and the way in which activity is framed (Cobb & Yackel, 1996). Despite heated disputes among the various camps or sects, the commonalties seem to be growing and some perspectives seem to complement each other rather than to clash (Billett, 1996; Cobb & Yackel, 1996; Ernest, 1999; Fosnot, 1996; Sfard, 1998; Vosniadou, 1996). Recently proponents of the cognitive and situative perspectives identified several important points on which they judge their perspectives to be in agreement (Anderson, Greeno, Reder, & Simon, 2000). Suggesting that the two approaches are different routes to the same goal, they declared that both perspectives "are fundamentally important in education...they can cast light on different aspects of the educational process" (p.11). A similar conclusion was reached by Cobb (1994) who claimed that each of the two perspectives "tells half of a good story" (p. 17). Another attempt at reconciling the two perspectives, yet from a different stance, was recently made by Packer and Goicoechea (2000). They argue that sociocultural and constructivist perspectives on learning presume different, and incommensurate, ontological assumptions. According to their claim what socioculturists call learning is the process of human change and transformation whereas what constructivists call learning is only part of that larger process. Yet they state that "whether one attaches the label "learning" to the part or to the whole, acquiring knowledge and expertise always entails participation in relationship and community and transformation both of the person and of the social world" (p.237).

Adhering to the reconciliatory stance, the following is an eclectic set of principles of learning and insights, distilled from the various competing perspectives, which looks to both schools of thoughts-- the individual and the social, though with a slight bias towards the latter.

Learning as active construction
- Learning is an active construction of meaning by the learner. (Meaning cannot be transmitted by direct instruction.)
- Discovery is a fundamental component of learning.
- For learning to occur, the learner has to activate prior knowledge, to relate new information/experience to it and restructure it accordingly.
- Learning is strategic. It involves the employment of cognitive and metacognitive strategies. (Self-regulated learners develop an awareness about when and how to apply strategies and to use skills, they monitor their learning process, evaluate and adjust their strategies accordingly.)
- Reflection is essential for meaningful learning.
- Learning is facilitated when: the student participates in the learning process and has control over its nature and direction about when and

how to apply strategies and to use skills, they monitor their learning process, evaluate and adjust their strategies accordingly.
- Reflection is essential for meaningful learning.
- Learning is facilitated when the student participates in the learning process and has control over its nature and direction.

Learning as a social phenomenon
- Learning is fundamentally social and derives from interactions with others (mind is distributed in society.) Cognitive change results from internalising and mentally transforming what is encountered in such interactions.

Learning as context related
- Learning is situated in a socio-cultural context. (What one learns is socially and culturally determined). Both social and individual psychological activity are influenced or mediated by the tools and signs in one's socio-cultural milieu.

Learning as participation
- Learning involves a process of enculturation into an established community of practice by means of cognitive apprenticeship.
- "Expertise" in a field of study develops not just by accumulating information, but also by adopting the principled and coherent ways of thinking, reasoning, and of representing problems shared by the members of the relevant community of practice.

Learning as influenced by motivation, affect and cognitive styles/intelligences
- What is constructed from a learning encounter is also influenced by the learner's motivation and affect: his/her goal orientation, expectations, the value s/he attributes to the learning task, and how s/he feels about it.
- Learning can be approached using different learning styles and various profiles of intelligences.

Learning as labour intensive engagement
- The learning of complex knowledge and skills requires extended effort and guided practice.

This mix of tenets represents a view that learning is a process of both self-organization and enculturation. Both processes take place as the learner participates in the culture and in doing so interacts with other participants. This view includes both the metaphor of "acquisition" and the metaphor of "participation" forwarded by Sfard (1998) for expressing prevalent frameworks of learning, which, despite and because of, their clashing definitions of learning, are both needed to explain that complex phenomenon.

2.2.2 Current Perspectives on Instruction

Contemporary definitions of good teaching emphasize its central function in facilitating students' learning. For instance, Biggs (1999) defines good teaching as "getting most students to use the high cognitive level processes that more academic students use spontaneously" (p. 73). Fenstermacher (1986) states that "the central task of teaching is to enable the student to perform the tasks of learning" (p 39). In order to facilitate learning, as it is conceptualised in constructivist frameworks, a paradigm shift in instruction, from *teaching-focused to learning-focused*, is essential. Central to this paradigm are concepts of autonomy, mutual reciprocity, social interaction and empowerment (Fosnot, 1996).

The role of the teacher under such a paradigm changes from that of an authoritative source of knowledge, who transmits this knowledge in hierarchically ordered bits and pieces, to that of a mentor or facilitator of learning who monitors for deep understanding. Likewise, the role of the student changes from being a passive consumer of knowledge to an active constructor of meaning. An important feature of the teaching-learning process is the dialogue between the teacher and the students through which, according to Freire (1972), the two parties "become jointly responsible for the process in which all grow" (p.53).

Biggs (1999) defines instruction as "a construction site on which students build on what they already know". (p. 72). The teacher, being the manager of this "construction site", assumes various responsibilities depending on the objectives of instruction and the specific needs that arise in the course of this process of construction. These responsibilities include: supervising, directing, counselling, apprenticing, and participating in a knowledge building community.

The learning environment that leads to conceptual development and change is rich in meaning-making and social negotiation. In such an environment, students are engaged in activity, reflection and conversation (Fosnot, 1996). They are encouraged to ask questions, explore, conduct inquiries; they are required to hypothesize, to suggest multiple solutions to problems; to generate conceptual connections, metaphors, personal insights, to reflect, justify, articulate ideas, elaborate, explain, clarify, criticize, etc. The learning tasks are authentic and challenging thus stimulating intrinsic motivation and fostering student initiative and creativity. Students are offered choice and control over the learning tasks and the classroom ethos is marked by a joint responsibility for learning. In this culture, hard work is valued and not perceived as a sign of a lack of ability.

An instructional approach that incorporates such features is problem-based learning (PBL). Briefly stated, PBL is a total approach to learning in which knowledge is acquired in a working context and is put back to use in that context. The starting point for learning is an authentic problem posed to the student, who needs to seek discipline-specific knowledge in order to solve it. The problems thus define what is to be learnt. Biggs (1999) argues "PBL is alignment itself. The objectives stipulate the problems to be solved, the main TLA [teaching-learning activity] is solving them, and the assessment is seeing how well they have been solved" (p. 207). He distinguishes five goals of PBL: (a) structuring functional knowledge; (b) developing effective professional reasoning processes; (c) developing self-regulated learning skills, (d) developing collaborative skills and (e) increasing motivation for learning. Instead of content coverage, students in PBL settings learn the skills for seeking out the required knowledge when needed. They are required to base decisions on knowledge, to hypothesize, to justify, to evaluate and to reformulate – all of which are the kind of cognitive activity that is required in current professional practice.

Emerging technologies of computer supported collaborative learning (CSCL) provide increasing opportunities for fostering learning in such an environment by creating on-line communities of learners. Computer mediated communication (CMC) is one such technology which enables electronic conferencing (Harasim, 1989). It offers a dynamic collaborative environment in which learners can interact, engage in critical thinking, share ideas, defend and challenge each others' assumptions, reflect on the learning material, ask questions, articulate their views, test their interpretations and synthesis, and revise and reconstruct their ideas. By fostering intersubjectivity among learners, this technology can thus help them negotiate meaning, perceive multiple problem-solving perspectives and construct new knowledge (Bonk & King, 1998; McLoughlin & Luca, 2000).

It is well acknowledged that successful implementation of such pedagogy entails, for many teachers, a radical change in their beliefs about knowledge, knowing and the nature of intelligence as well as in their conceptions of learning and teaching.

Form a theory-of-mind perspective, Olson and Bruner (1997) present two models of learners' minds that bear close resemblance to the perspective of this constructivist-based pedagogy. One of these models conceptualises the learner as a thinker and the other as an expert. The folk psychology regarding the former conceives learners as being able to understand, to reason, to reflect on their ideas, to evaluate them, and correct them when needed. Learners, it is claimed, have a point of view, hold more or less coherent "theories" about the world and the mind, and can turn beliefs into hypotheses to be openly tested. The learner is conceived of as an interpreter who is engaged in constructing a model of the world. The corresponding folk pedagogy conceives the teacher's role as that of a collaborator who tries to understand what the learners think and how they got there. The learning process features a dialogue -- an exchange of understanding between teacher and learner. The folk psychology regarding the other model – the learner as an expert –conceives of the learner's mind as an active processor of beliefs and theories that are formed and revised based on evidence. The learner, so it is claimed, recognizes the distinction between personal and cultural knowledge. Learning is of the peer collaboration type whereby the learner assumes the role of knowledge co-constructor. The corresponding folk pedagogy conceives the teacher's role as an information manager who assists the learners in evaluating their beliefs and theories reflectively and collaboratively in light of evidence and cultural knowledge. Olson and Bruner classify these two models as internalist theories of mind, stating that unlike the externalist theories that focus on what the teacher can do to foster learning, internalist theories focus "on what the learners can do or what they think they are doing, and how learning can be premised on those intentional states" (p.25). They further argue that internalist theories aspire to apply the same theories to learners as learners apply to themselves, as opposed to the objective, detached view espoused by externalist theories.

2.2.3 Current Perspective on Assessment

The assessment approach that is aligned with the constructivist-based teaching approach is sometimes referred to as *assessment culture*, as opposed to the conservative *testing culture* (Birenbaum, 1996; Gipps, 1994; Wolf, Bixby, Glenn, & Gardner, 1991). While the conservative approach reflected a psychometric-quantitative paradigm, the constructivist approach reflects a contextual-qualitative paradigm. This approach strongly emphasizes the integration of assessment and instruction and focuses on the assessment of the process of learning in addition to that of its products. The assessment itself takes many forms, all of which are generally referred to by psychometricians as "unstandardized assessments embedded in instruction" (Koretz, Stecher, Klein, & McCaffrey, 1994). Reporting practices shift from single total scores, used in the testing culture for ranking students, to descriptive profiles that provide multidimensional feedback for fostering learning. In this culture the position of the student with regard to the assessment process changes from that of a passive, powerless, often oppressed, subject who is mystified by the process, to being an active participant who shares responsibility in the process. Students participate in the development of the criteria and the standards for evaluating their own performance, they practice self- and peer-assessment and are required to reflect on their learning and to keep track of their academic growth.

These features of the assessment culture make it most suitable for formative classroom assessment, which is geared to promote learning, as opposed to summative high-stakes (often large-scale) assessment that serves accountability as well as certification and selection purposes (Koretz, et al., 1994; Worthen, 1993).

Feedback has always been at the heart of formative assessment. Metaphorically speaking, if we liken alignment of instruction, learning and assessment to a spin top then feedback is the force that spins the top. Feedback as a general term has been defined as "information about the gap between the actual level and the reference level of a system parameter which is used to alter the gap in some way" (Ramaprasad, 1983, p.3). The implications for assessment are that in order for it to be formative, the information contained in feedback must be of high quality and mindfully used. This entails that the learner first realizes the gap between the desired goal (the reference) and his/her current level of understanding and identifies the causes of this gap, and then acts to close it (Black & Wiliam, 1998; Ramaprasad, 1983; Sadler, 1989). Teachers and computer-based instructional systems were the providers of feedback in the past. The new perspective on assessment stresses the active role of the learner in generating feedback. Self- and peer- assessment is therefore highly recommended for advancing understanding and promoting self-regulated life long learners (Black & Wiliam, 1998).

New forms of assessment

The main tools employed in the assessment culture for collecting evidence about learning are performance tasks, learning journals and portfolios. These tools are well known to the readers and therefore they will only be briefly presented, just for the sake of completeness. Unlike most multiple-choice items, *performance tasks* are designed to tap higher order thinking such as planning, hypothesizing, organizing, integrating, criticizing, drawing conclusions, evaluating, etc.; they are meant to elicit what Perkins and Blythe (1994) term "understanding performances". Typically performance on such tasks is not subject to time limitations, and a variety of tools, including those used in real life for performing similar tasks, are permitted. The tasks are complex, often refer to multidisciplinary contents, they have more than one single possible solution or solution path, and are loosely structured. This requires the student to identify and clearly state the problem. The tasks, typically involving investigations of various types, are meaningful and authentic to the practice in the discipline, and they aim to be interesting, challenging and engaging for the students, who often perform them in teams. Upon completion of the task, students are frequently required to exhibit their understanding in a communicative manner. Analytic or holistic rubrics that specify clear benchmarks of performance at various levels of proficiency serve the dual purpose of guiding the students as they perform the task as well as guiding the raters who evaluate the performance. They are also used for self- and peer-assessment.

Learning journals are used for documenting students' reflections on the material and their learning processes. The learning journal thus promotes the construction of meaning as well as contributes valuable evidence for assessment purposes. Learning journals can be used to assess the quality of knowledge (Sarig, 1996) and the learner's reflective and metacognitive competencies (Birenbaum & Amdur, 1999).

The *portfolio* best serves the dual purpose of learning and assessment. It is a container that holds a purposeful collection of evidence of the student's learning efforts, process, progress, and accomplishments in (a) given area(s). When implementing portfolios it is essential that the students participate in the selection of its content, of the guidelines as well as of the criteria for assessment (Arter & Spandel, 1992; Birenbaum, 1996). When integrated, evidence collected by means of these tools can provide a comprehensive and realistic picture of what the student knows and is able to do in a given area.

2.2.4 Relationships between Assessment and Learning

Formative assessment is expected to improve learning. It can meet this expectation and indeed was found to do so (Black & Wiliam, 1998) but this is not a simple process and occasionally it fails. What are the factors that might interfere with the process and cause its failure? In order to answer this question let's examine the stages a learner proceeds through from the moment s/he is faced with an assessment task until s/he reaches a decision as to how to respond to the feedback information. Below are listed the stages and what the learner needs to posses/know/do with respect to each of them.

Stage I: Getting the task
- Interpret the task in congruence with the teacher's intended goals.
- Understand the task requirements and the standard that it is addressing.
- Have a clear idea of what an outcome that meets the standard looks like.
- Know what strategies should be applied in order to successfully perform the task.

- Perceive the value of the task and be motivated to perform it to the best of his/her capability.
- Be confident in his/her capability to perform the task successfully (self-efficacy).

Stage II: Performing the task
- Effectively apply the relevant cognitive strategies.
- Effectively apply metacognitive strategies to monitor and regulate performance.
- Effectively manage time and other relevant resources.
- Effectively control and regulate his/her feelings.
- If given a rubric, appropriately interpret its benchmarks.
- Be determined to invest the necessary efforts to complete the task properly.

Stage III: Appraising performance and generating feedback information
- Accurately asses (by him/herself or with the help of the teacher and/or peers) his/her performance.
- In case of a gap between the actual performance and the standard – understand the goals he/she is failing to attain.
- Understand what caused the failure.
- Conceive of mistakes as a springboard toward growth rather than just a sign of low ability; consequently, not attaining the goals does not affect his/her self-image and self-efficacy.
- Posses a mastery orientation towards learning.
- Feel committed to close the gap.
- State self-referenced goals for pursuing further learning in order to close the gap.

Learners vary in their profile with respect to the features listed above, and consequently formative assessment occasionally fails to improve learning. Yet, classroom ethos and other features of the learning environment, as well as teachers' and students' beliefs about knowledge, learning and teaching can reduce this variance thus affecting the rate of success of formative assessment (Birenbaum, 2000; Black & Wiliam, 1998).

To conclude the theoretical framework, here are the main attributes of an aligned ILA system that is suitable for achieving the goals of higher education in the knowledge age: Instruction-wise: learning focused; learning-wise: reflective-active knowledge construction; assessment-wise: contextualized, interpretative and performance-based.

3. FROM THEORY TO PRACTICE: AN ILA CULTURE

This part briefly describes an ILA culture based on constructivist principles that was created in a graduate course dedicated to alternatives in assessment taught by the author. It illustrates the application of methods and tools such as those discussed earlier in this chapter and exemplifies that instruction, learning and assessment are inextricably bound up with each other, making up a whole that is more than the sum of its parts.

3.1 Aims of the Course

The ILA culture developed in this two-semester course is aimed to introduce the students (most of whom are in-service educators – teachers, principals, superintendents -- who pursue their master's or doctoral studies) to a culture that is conducive to the implementation of alternative assessment. It offers the students an opportunity, through personal experience, to deepen their understanding regarding the nature of this type of assessment and concerning its role as an integral part of the teaching-learning process. At the same time, it offers them a chance to develop their own reflective and other self-regulated learning capabilities. Such capabilities are expected to support their present and future learning processes (Schunk & Zimmerman, 1994) as well as their professional practice (Schön, 1983).

3.2 Design Features

The course design is rooted in constructivist notions about knowledge and knowing and the derived conceptions of learning and teaching, and it is geared to elicit individual and social knowledge construction through dialogue and reflection. It uses a virtual environment that complements the regular classroom meetings to create a knowledge building community by means of asynchronous electronic discussion forums (e-forums). Included in this community are also students who took the course in previous years who have chosen to remain active members of the knowledge building community. Recently the learning environment has been augmented to offer enrichment materials. In its current form it includes short presentations of various relevant topics with links to references, video clips, power-point presentations, Internet resources, examples of learning outcomes from previous years, etc. Information is cross-referenced and can be retrieved through various links. The learning environment is accessed through a "city map" which is meant to visually orient the learners regarding the scope and structure (relationships among concepts) of the assessment domain and its intersections with the domains of learning and instruction. Icons that resemble public institutions and other functional artefacts of a city help the learner navigate while exploring the "ILA City".

The features of the culture created in the course include: freedom of choice, openness, flexibility, student responsibility, student participation, knowledge sharing, responsiveness, support, caring, empowerment, and mutual respect.

3.3 Instruction

The instruction is learning-focused. Central features of the pedagogy are dialogue, reflection and participation. The instructor facilitates students' learning by engaging them in discussions, conversations, collaborative inquiry projects, experiments, authentic performance tasks and reflection, as well as by modelling good and bad practice illustrated by means of her own and other professionals' behaviour, through video tapes and field trips. Feedback and guidance are regularly provided to students as they work on their projects. The instructional process is responsive to students' needs and interests. The discussions centre on authentic issues and dilemmas concerning assessment but there is no fixed set of topics nor pre-specified sequencing. There is no anticipation that all students leave the course with the same knowledge base. Rather each student is expected to deepen his/her understanding of the aspects that are most relevant to his/her professional practice and be able to make educated decisions regarding assessment on the basis of the knowledge constructed during the course.

3.4 Learning

The learning is reflective-active. Students are engaged in personal-and group-knowledge construction by means of group projects, discussions held in class and in the e-forums, and learning journals in which they reflect on what has been learnt in class and from the assigned reading materials. In the e-forums, students share with the other community members reflections they have recorded in their learning journals, and they record back in their journals the insights gained from the discussion. Students work on their projects in teams meeting face-to-face and/or through sub-forums opened for each project at the course's virtual site. Towards the end of the course they present their project outcomes in a plenary session and receive written feedback from their peers and the instructor. They then use this feedback for revising their work and are required to hand in a written response to the feedback along with the final version of their work. During the course students study the textbook and assigned papers and they retrieve relevant materials for their projects form various other resources. In addition, each student is nominated a "web-site reporter" which entails frequent visits to a given internet site dedicated to assessment and reporting back to class when relevant information regarding the topic under discussion is retrieved.

3.5 Assessment

The assessment is performance based, integrated and contextualized. It serves both formative and summative purposes. The following three products are assessed providing a variety of evidence regarding the learning outcomes:

- *Retrospective summary of learning* - At course end, students are required to review their on-going journal entries and their postings in the e-forums and prepare a retrospective summary. The summary is meant to convey their understanding regarding the various aspects of assessment addressed in the course and their awareness of their personal growth with respect to this domain. The holistic rubric, jointly developed with the students, comprises the following criteria: quality of knowledge (veritability, complexity, applicability), learning disposition (critical, motivated), self-awareness of progress, and contribution to the knowledge building community.
- *Performdnce assessment project.* Students are required to develop an assessment task, preferably an interdisciplinary one. This involves the definition of goals for the assessment, formulation of the task, development of a rubric for assessing performance, administration of the task, analysis of the results, and generation of feedback as well as critical evaluation of the quality of the assessment.
- *"Position paper".* This project is introduced as a collective task in which the class writes a position paper to the Ministry of Education regarding the constructivist-based assessment culture. (It should be noted that this is an authentic task given that interest in the new forms of assessment is recent among Israeli policy makers.) Students propose controversial issues relevant to the field and each team chooses an issue to study and write about. The jointly developed analytic rubric for assessing performance on this task consists of the following dimensions: issue definition and importance, content, sources of information, argument, conclusions and implications, and communicability with emphasis on audience awareness.

Students work on their projects throughout the course. The features of the assessment process are as follows:

- On-going feedback - provided throughout the course by the instructor in accordance with each student/group's particular needs. This feedback loop is conducted both through the project e-forum and through face-to-face meetings. The other community members have access to the project e-forum and are invited to provide feedback or suggestions.
- Student participation – Students take part in the decision making process regarding the assessment. They participate in the development of the assessment rubrics and have a say in how their final grades are to be weighted.
- Self- and peer-assessment. The same rubric is used by the instructor and by the students. The latter use it to assess self- and peer- performance. After the students submit their work for final assessment, including their self-assessment, the instructor provides each student with detailed feedback and meets the student for an assessment conference if a discrepancy between the assessments occurs.

As to course evaluation, students' average rating of the course is quite high and their written comments indicate that they consider it a very demanding yet a profound learning experience. The same can be said from the standpoint of the instructor.

4. DIRECTIONS FOR RESEARCH ON ASSESSMENT FOR LEARNING

Research evidence conclusively shows that assessment *for* learning improves learning. Following a thorough literature review about assessment and classroom learning Black and Wiliam (1998) conclude that the gains in achievement due to formative assessment "appear to be quit considerable... among the largest ever reported for educational interventions" (p. 61). They note that these gains were evident where innovations designed to strengthen the frequent feedback about learning were implemented. However, Black and Wiliam also claim "it is clear that most of the studies in the literature have not attended to some of the important aspects of the situations being researched" (p.58). Stressing that "the assessment processes are, at heart, social processes, taking place in social settings, conducted by, on and for social actors" (p.56) they point to the absence of contextual aspects from much of the research they reviewed. In other words, the ILA culture in which assessment for learning is embedded has yet to be empirically investigated. Figure 1 displays some context-related constructs subsumed under the ILA culture and their hypothesized interrelationships and impact on assessment. Although this mapping by no means captures the entire network of relevant constructs and interrelationships it suffices to illustrate the intricacy of such a network. Represented in the mapping are constructs related to class regime and climate; learning environment; teachers' and learners' epistemological beliefs, conceptions of learning, teaching and assessment; teachers' knowledge, skills, and strategies; learners' motivation, competencies and strategies; learning interactions and consequent knowledge construction; and finally, assessment strategies and techniques with special emphasis on feedback.

As can be seen in the figure the hypothesized relationships among these constructs create a complex network of direct, indirect and reciprocal effects. The question, then, arises as to how this intricacy can be investigated. Judging by the nature of the key constructs a quantitative approach, employing even sophisticated multivariate analyses such as SEM (structural equation modelling), will not suffice. A qualitative approach seems a better choice. Cultures are commonly studied by means of ethnographic methods, but even among those, the conventional ones may not be sufficient. Eisenhart (2001) has recently criticized the sole reliance on conventional ethnographic methods arguing that such methods and ways of thinking about and looking at cultures are not enough if we want to grasp the new forms of life, including school culture, in the post-modern era. She notes that these forms "seem to be faster paced, more diverse, more complicated, more entangled than before" (p. 24). Research aimed at understanding the ILA culture will therefore need to incorporate a variety of conventional and non-conventional ethnographic methods and perspectives that fit the conditions and experiences of such culture.

New Insights Into Learning and Teaching 31

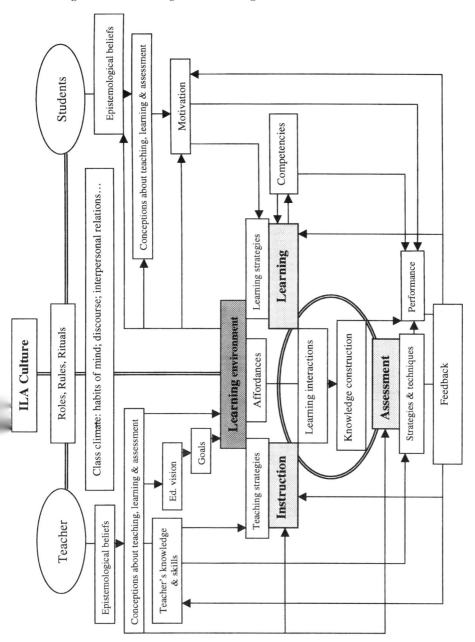

Figure 1. Mapping of Constructs Related to the ILA Culture

Further research is also needed regarding the nature of assessment-related constructs in light of recent shifts in their conceptualisation. For instance, feedback is currently conceptualised as a joint responsibility of the teacher and the learner (Black & Wiliam, 1998). Consequently, further research is needed to better understand the nature of self- and peer-assessment and their impact on learning. Relevant research questions might refer to the process whereby the assessment criteria are being negotiated; to how students come to internalise the standards for good performance; to the accuracy of self- and peer-assessments; to how they affect learners' self-efficacy and other related motivational factors, etc. Another related construct whose underlying structure deserves further research is conceptions of assessment. Relevant research questions might refer to the nature of teachers' and students' conceptions of good assessment practice and their respective roles in the assessment process; to the effect of the learning environment on students' conceptions of assessment; to the relationships between teachers' conceptions of assessment, their mental models and their interpretations and use of evidence of learning, etc. It is obvious that in order to answer such questions a variety of quantitative and qualitative methods will have to be employed.

Assessment of learning interactions in a virtual learning environment is yet another area in which further research is needed due to the accelerated dispersion of distance learning in higher education. Questions such as: How to efficiently present feedback information during on-line problem solving? How to assess students' contribution to the knowledge building community in an e-discussion forum? etc. are examples of a wide variety of timely, practical assessment-related questions that need to be properly addressed.

Another line of research is related to teacher training in assessment. It is well acknowledged that most teachers currently employed have not been systematically trained in assessment either in teacher preparation programs or in professional development while on the job (Popam, 2001). The situation is even more acute in higher education where most instructors do not receive systematic pedagogical training of any kind. Since their majority left school before the new forms of assessment were introduced they have never been exposed to this type of assessment. In view of this situation, research should be directed at designing effective training interventions tailored to the needs of these populations.

The issues addressed so far relate to formative assessment; however, for certain purposes, such as certifying and licensing, there is a need for high-stake summative assessment. Quality control issues then become crucial. They refer to the accuracy (reliability) of the assessment scores as well as to the validity of the inferences drawn on the basis of these scores. Research has shown that the new forms of assessment tend to compare unfavourably to standardized testing with respect to these psychometric criteria of reliability and validity (Dunbar, Koretz, & Hoover, 1991; Koretz, Stecher, Klein, & McCaffrey, 1994; Linn, 1994). Consequently, standardized tests are mostly used for high-stake summative assessment. For practical purposes this dichotomy between summative and formative assessment is problematic. Further efforts should therefore be made to conceptualise more suitable criteria for quality control with respect to the new forms of assessment. This direction complies with criticism raised regarding the applicability of psychometric models to this type of assessment (Birenbaum, 1996; Delandshere & Petrosky, 1994; 1998; Dierick & Dochy, 2001; Moss, 1994, 1996) and in general to the context of classroom assessment (Dochy & Moerkerke, 1997). On this line, principles of an edumetric approach have recently been suggested (Dierick & Dochy, 2001) that expand the traditional concepts of validity and reliability to include assessment criteria that are sensitive to the intricacy of the teaching-learning process. The operationalization of these criteria and their applicability will need to further be investigated, along with the impact of their implementation.

In conclusion, it seems that the assessment community has come a long way since the new forms of assessment were introduced more than a decade ago. It has deepened its understanding regarding the role and potential of assessment in the instruction-learning process and its context. These understandings, together with the new intriguing options provided by ICT (information communication technology), have opened up new horizons for research on methods for optimising assessment in the service of learning. Embarking on these lines of research will undoubtedly contribute significantly to the joint efforts to meet the challenge for higher education in the knowledge age.

REFERENCES

Anderson, J. R., Greeno, J. G., Reder, L. M., & Simon H. A. (2000). Perspectives on learning, thinking, and activity. *Educational Researcher, 29* (4), 11-13.

Arter, J. A., & Spandel, V. (1992). Using portfolios of student work in instruction and assessment. *Educational, Measurement: Issues and Practice, 11* (1), 36-44.

Atkins, M. (1995). What should we be assessing? In P. K. Knight (Ed.), *Assessment for learning in higher education* (pp.24-33). London: Kogan Page.

Bell, D. (1999). *The coming of post-industrial society.* New York: Basic Books.

Biggs, J. (1999). *Teaching for quality learning at university.* Buckingham: The Society for Research into Higher Education & Open University Press.

Billett, S. (1996). Situated learning: Bridging sociocultural and cognitive theorizing. *Learning and Instruction, 6* (3), 263-280.

Birenbaum, M. (1996). Assessment 2000: Toward a pluralistic approach to assessment. In M. Birenbaum & F.J.R.C. Dochy (Eds.), *Alternatives in assessment of achievement, learning processes and prior knowledge* (pp. 3-29). Boston, MA: Kluwer.

Birenbaum, M. (Sept. 13, 2000). *New insights into learning and teaching and the implications for assessment.* Keynote address at First Conference of the Special Interest Group on

Assessment and Evaluation of the European Association for Research on Learning and Instruction (EARLI). Maastricht, The Netherlands.

Birenbaum, M., & Amdur, L. (1999). Reflective active learning in a graduate course on assessment. *Higher Education Research and Development, 18* (2), 201-218.

Black, P., & Wiliam, D. (1998). Assessment and classroom learning. *Assessment in Education, 5* (1) 7-74.

Blumer, H. (1969) *Symbolic interactionism: Perspectives and method.* Englewood Cliffs, NJ: Prentice-Hall.

Bonk, C. J., & King, K. S. (Eds.). (1998). *Electronic collaborators: Learner-centered technologies for literacy, apprenticeship, and discourse.* Mahwah, NJ: Erlbaum.

Brown, J. S., Collins, A., & Duguid, P. (1989). Situated cognition and the culture of learning. *Educational Researcher, 18* (1), 32-42.

Cobb, P. (1994). Where is the mind? Constructivist and sociocultural perspective on mathematical development. *Educational Researcher, 23* (7), 13-20.

Cobb, P., & Yackel, E. (1996). Constructivist, emergent, and sociocultural perspectives in the context of developmental research. *Educational Psychologist, 31* (3/4), 175-190.

Cunningham, J. W., & Fitzgerald, J. (1996). Epistemology and reading. *Reading Research Quarterly, 31* (1), 36-60.

Delandshere, G., & Petrosky, A. R. (1994). Capturing teachers' knowledge: Performance assessment. *Educational Researcher, 23* (5), 11-18.

Delandshere, G., & Petrosky, A. R. (1998). Assessment of complex performances: Limitations of key measurement assumptions. *Educational Researcher, 27* (2), 14-24.

Derry, S. J. (1996). Cognitive schema theory in the constructivist debate. *Educational Psychologist, 31* (3/4), 163-174.

Dewey, J. (1981). Experience and nature. In J. A. Boyston (Ed.), *John Dewey: The later works, 1925-1953*, Vol. 1. Carbondale: Southern Illinois University Press. (Original work published 1925).

Dierick, S., & Dochy, F. (2001). New lines in edumetrics: new forms of assessment lead to new assessment criteria. *Studies in Educational Evaluation, 27*, 307-329

Dochy, F., & Moerkerke, G. (1997). Assessment as a major influence on learning and instruction. *International Journal of Educational Research, 27* (5), 415-432,

Dunbar, S. B., Koretz, D. M., & Hoover, H. D. (1991). Quality control in the development and use of performance assessments. *Applied Measurement in Education, 4*, 289-303.

Eisenhart, M. (2001). Educational ethnography, past, present, and future: Ideas to think with. *Educational Researcher, 30* (8), 16-27.

Entwistle, N. J., & Entwistle, A. (1991). Constructing forms of understanding for degree examination: the student experience and its implications. *Higher Education, 22*, 205-227.

Ernest, P. (1999). Forms of knowledge in mathematics and mathematics education: Philosophical and rhetorical perspectives. *Educational Studies in Mathematics, 38*, 67-83.

Fenstermacher, G. D. (1986). Philosophy of research on teaching: Three aspects. In M. C. Wittrock (Ed.), *Handbook of research on teaching (3rd edition)* (pp. 37-49). New York: Macmillan.

Feuerstein, R. (1980). *Instrumental enrichment: An intervention program for cognitive modifiability.* Baltimore, MD: University Park Press.

Fosnot, C. T. (1996). Constructivism: A psychological theory of learning. In C. T. Fosnot (Ed.), *Constructivism: Theory, perspectives, and practice.* New York: Teachers College Press.

Freire, P. (1972). *Pedagogy of the oppressed.* (M. Bergman Ramos trans.) Harmondsworth, UK: Penguin

Gardner, H. (1983). *Frames of mind: The theory of multiple intelligences.* New York: Basic Books.
Gardner, H. (1993). *Multiple intelligences: The theory in practice.* New York: Basic Books.
Gergen, K. J. (1994). *Realities and relationships. Soundings in social construction.* Cambridge, MA: Harvard University Press.
Gipps, C. (1994). *Beyond testing: towards a theory of educational assessment.* London: Palmer Press.
Glatthorn, A. A., & Jailall, J. (2000). Curriculum for the new millenium. In R. S. Brandt (Ed.), *Education in a new era* (pp. 97-121). Alexandria, VA: Association for Supervision and Curriculum Development (ASCD).
Guba, E. G. (Ed.). (1990). *The paradigm dialogue* (pp. 17-42). Thousand Oaks, CA: Sage.
Harasim, L. (1989). Online education. A new domain. In R. Mason & A. Kaye (Eds.), Mindweave: *Communication, computers and distance education* (pp. 50-62). Oxford: Pergamon.
Jarvis, P., Holford, J., & Griffin, C. (1998). *The theory and practice of learning.* London: Kogan Page.
Koretz, D., Stecher, B., Klein, S, & McCaffrey, D. (1994). The Vermont portfolio assessment program: Findings and implications. *Educational Measurement; Issues and Practice, 13* (3), 5-16.
Lave, J., & Wenger, E. (1991). *Situated learning: Legitimate peripheral participation.* Cambridge University Press.
Linn, R. L. (1994). Performance assessment: Policy promises and technical measurement standards. *Educational Researcher, 23* (9), 4-14.
Mayer, R. E. (1996). Learners as information processors: Legacies and limitations of educational psychology's second metaphor. *Educational Psychologist, 31* (3/4), 151-161.
McLoughlin, C., & Luca, J. (2000). Cognitive engagement and higher order thinking through computer conferencing: We know why but do we know how? In A. Herrmann and M. M. Kulski (Eds.), *Flexible futures in tertiary teaching.* Proceedings of the 9th Annual Teaching Learning Forum, 2-4 February 2000. Perth: Curtin University of Technology. Available: http://cleo.murdoch.edu.au/confs/tlf/tlf2000/mcloughlin.html
Moss, P. A. (1994). Can there be validity without reliability? *Educational Researcher, 23* (2) 5-12.
Moss, P. A. (1996). Enlarging the dialogue in educational measurement: Voice from interpretive research traditions. *Educational Researcher, 24* (1), 20-28, 43.
Olson, D. R., & Bruner, J. S. (1997). Folk psychology and folk pedagogy. In D. R. Olson & N. Torrance (Eds.), *Handbook of education and human development* (pp. 9-27). London: Blackwell.
Packer, M. J., & Goicoechea, J. (2000). Sociocultural and constructivist theories of learning: ontology, not just epistemology. *Educational Psychologist, 35* (4), 227-242.
Perkins, D. N., & Blythe, T. (1994). Putting understanding up front. *Educational Leadership, 51* (5), 11-13.
Phillips, D. C. (1995). The good, the bad, and the ugly: The many faces of constructivism. *Educational Researcher, 24* (7), 5-12.
Pintrich, P. R. (2000). The role of orientation in self-regulated learning. In M. Boekaerts, P. R. Pintrich, & M. Zeidner, (Eds.), *Handbook of self-regulation.* (pp. 451-502). San Diego: Academic Press.
Popam, W. J. (2001). *The truth about testing.* Alexandria, VA: Association for Supervision and Curriculum Development (ASCD).

Prawat, R. S. (1996). Constructivism, modern and postmodern. *Educational Psychologist, 31* (3/4), 215-225.

Ramaprasad, A. (1983). On the definition of feedback. *Behavioral Science, 28*, 4-13.

Ramsden, P. (1987). Improving teaching and earning in higher education: the case for a relational perspective. *Studies in Higher Education, 12*, 275-286.

Randi, J., & Corno, L. (2000). Teacher innovations in self-regulated learning. In M. Boekaerts, P. R. Pintrich, & M. Zeidner, (Eds.), *Handbook of self-regulation*. (pp. 651-685). San Diego: Academic Press.

Sadler, R. (1989). Formative assessment and the design of instructional systems. *Instructional Science, 18*, 119-144.

Sarig, G. (1996). Academic literacy as ways of getting to know: What can be assessed? In M. Birenbaum, & F. J. R. C. Dochy (Eds.), *Alternatives in assessment of achievements, learning processes and prior knowledge* (pp.161-199). Boston, MA: Kluwer.

Schön, D. A. (1983). *The reflective practitioner: How professionals think in action*. New York: Basic Books.

Schunk, D. H., & Zimmerman, B. J. (Eds.). (1994). *Self-regulation of learning and performance*. Hillsdale, NJ: Erlbaum.

Sfard, A. (1998). On two metaphors for learning and the dangers of choosing just one. *Educational Researcher, 27* (2), 4-13.

Sternberg, R. J. (1985). *Beyond IQ: A triarchic theory of human intelligence*. New York: Cambridge University Press.

Sternberg, R. J. (1986). *The triarchic mind: A new theory of human intelligence*. NY: Viking

Tynjälä, P. (1999). Towards expert knowledge? A comparison between a constructivist and a traditional learning environment in the university. *International Journal of Educational Research, 31*, 357-442.

Von Glasersfeld, E. (1995). Sensory experience, abstraction, and teaching. In L. P. Steffe & J. Gale (Eds.), *Constructivism in education* (pp. 369-383). Hillsdale, NJ: Erlbaum.

Vosniadou, S. (1996). Towards a revised cognitive psychology for new advances in learning and instruction. *Learning and Instruction, 6* (2), 95-110.

Vygotsky, L. S. (1978). *Mind in society*. Cambridge, MA: Harvard University Press.

Wolf, D., Bixby, J., Glenn, J., & Gardner, H. (1991). To use their minds well: Investigating new forms of student assessment. *Review of Research in Education, 17*, 31-73.

Worthen, B. R. (1993). Critical issues that will determine the future of alternative assessment. *Phi Delta Kappan, 74*, 444-456.

Zimmerman, B. J. (1989). Models of self-regulated learning and academic achievement. In B. K. Zimmerman & D. H. Schunk (Eds.), *Self regulated learning and academic achievement: Theory, research, and practice* (pp. 1-25). New York: Springer-Verlag.

Evaluating the Consequential Validity of New Modes of Assessment: The Influence of Assessment on Learning, Including Pre-, Post-, and True Assessment Effects

Sarah Gielen[1], Filip Dochy[1] & Sabine Dierick[2]
[1]University of Leuven, Department of Instructional Science, Centre for Research on Teacher and Higher Education, Belgium, [2]University of Maastricht, The Netherlands

1. INTRODUCTION

The role of assessment and evaluation in education has been crucial, probably since the earliest approaches to formal education. However, much more attention has been paid to this role in the last few decades, largely due to wider developments in society. The most fundamental change in our views of assessment is represented by the notion of assessment as a tool for learning (Dochy & Mc Dowell, 1997). Whereas in the past, we have seen assessment primarily as a means to determine measures and thus for certification, there is now a belief that the potential benefits of assessing are much wider and impinge on in all stages of the learning process. The new assessment culture (Birenbaum & Dochy, 1996) strongly emphasises the integration of instruction and assessment. Students play far more active roles in the assessment of their achievement. The construction of tasks, the development of criteria for the assessment and the scoring of performance may be shared or negotiated among teachers and students. The assessment takes all kinds of forms such as observations, text- and curriculum-embedded questions and tests, interviews, performance assessments, writing samples, exhibitions, portfolio assessment, overall assessment. Several labels have been used to describe subsets of these assessment modes, with the most common being "direct assessment", "authentic assessment", "performance assessment" and "alternative assessment". It is widely accepted that these new forms of assessment lead to a number of benefits in terms of the learning process: encouraging thinking, increasing learning and also increasing students' confidence (Falchikov, 1986; 1995).

One could argue that a new assessment culture cannot be evaluated solely on the basis of pre-era criteria. To do right to the basic assumptions of these assessment forms, the traditionally used psychometric criteria need to be expanded, and additional relevant criteria for evaluating the quality of assessment need to be developed (Dierick & Dochy, 2001). In this respect, the concept "psychometrics" is often replaced by the concept of "edumetrics".

In this contribution, we will first focus on the criteria that we see as necessary to expand the traditional psychometric criteria to evaluate the quality of assessments. In a second part, we will outline some of the characteristics of new modes of assessment and relate these to their role within the consequential validity.

2. EVALUATING NEW MODES OF ASSESSMENT ACCORDING TO THE NEW EDUMETRIC APPROACH

Various authors have recently proposed ways to extend the criteria, techniques and methods used in traditional psychometrics in order to evaluate the quality of assessments. Within the literature on quality criteria for evaluating assessment, a difference can be made between authors, who present a more expanded vision on validity and reliability (Cronbach, 1989; Kane, 1992; Messick, 1989) and those who propose specific criteria, sensitive to the characteristics of new modes of assessment (Fredericksen & Collins, 1989; Haertel, 1991; Linn, Baker & Dunbar, 1991).

If we integrate the most important changes within the assessment field with regard to the criteria for evaluating assessment, conducting quality assessment inquiry involves a comprehensive strategy that addresses evaluating:
1. The validity of assessment tasks.
2. The validity of assessment scoring.
3. The generalizability of assessment.
4. The consequential validity of assessment.

During this inquiry, arguments will be found that support or refute the construct validity of assessment. Messick (1989) suggested that two questions must be asked whenever a decision about the quality of assessment is made.

First, is the assessment any good as a measure of the characteristics it is interpreted to assess? Second, should the assessment be used for the proposed purpose?

In evaluating the first question, evidence of the validity of the assessment tasks, of the assessment performance scoring and the generalizability of the assessment must be considered. The second question evaluates the adequacy of the proposed use (intended and unintended effects), against alternative means of serving the same purpose. In the evaluative argument, the evidence obtained during validity inquiry will be considered and carefully weighted, to reach a conclusion about the adequacy of assessment use for the specific purpose.

In table 1, an overview is given of questions that can be used as guidelines to collect supporting evidence for and examining possible threats to construct validity.

Arguments, supportive for the construct-validity of new assessment forms, will be outlined shortly below.

2.1   Validity of Assessment Tasks Used

Assessment development begins with establishing an explicit conceptual framework that describes the construct domain being assessed: content and cognitive specifications should be identified. During the first stage, validity inquiry judges how well assessment matches the content and cognitive specifications of the construct measured (Shavelson, 2002). The defined framework can be used as a guideline to select assessment tasks.

The following aspects are important to consider: First, the tasks used must be an appropriate reflection of the construct or, as one should perhaps say within the assessment culture, the competence that needs to be assessed. Secondly, with regard to the content, it means that the tasks should be authentic in that they are representative of the real life problems that occur in the knowledge domain assessed. Third, the cognitive level needs to be complex, so that the same thinking processes are required than experts use for solving domain-specific problems.

New assessment forms score better on these criteria than so-called objective tests, precisely because of their authentic and complex problem character (Dierick & Dochy, 2001).

Table 1. A Framework for Collecting Supporting Evidence for and Examining Possible Threats to Construct Validity

Procedure	Review questions regarding
1. Establish an explicit conceptual framework for the assessment = **Construct definition**: (content and cognitive specifications) *Purpose* Frame of reference for reviewing assessment tasks or items with regard to the purport construct/comptence	**1. VALIDITY OF THE ASSESSMENT TASKS** Judging how well assessment matches **the content and cognitive specifications** of the construct/competency that is measured ⇨ A. Consists assessment of a *representative set of tasks* that cover the spectrum of knowledge and skills needed for the construct/competence measured? B. Are the tasks *authentic* in that they are representative of the real life problems that occur within the construct/competence domain that is measured? C. Are the tasks assessing *complex abilities in that* the same thinking processes are required for solving the tasks than experts use for solving domain-specific problems.?
1. Identify rival explanations for the observed performance ⇨ Collect multiple types of evidence on the rival explanations of assessment performance *Purpose* Identifies possible weaknesses in the interpretation of scoring ⇨ Provides a basis for refuting alternative explanations for performance or for revising the assessment, if necessary.	**2. VALIDITY OF ASSESSMENT SCORING** Searching evidence for the appropriateness of the inference from *the performance to an observed score* ⇨ A. Is the task congruent with the received instruction/education? B. Do all students have equal opportunities to demonstrate their capabilities on basis of the selected tasks? - Can all students demonstrate their performance in the selected mode of assessment? - Are students accustomed with the cultural content that is asked for? - Are students enough motivated to perform well? - Can students use the necessary equipment ? - Do students not enjoy inappropriate advantages ? A. Do the scoring criteria relate directly to the particular construct/competence that the task is intended to measure? B. Are the scoring criteria clearly defined (transparent)? C. Do students understand the criteria that are used to evaluate their performance?

3. THE GENERALIZABILITY OF THE ASSESSMENT SCORE

Searching for evidence of the appropriateness of the inference from the observed score to a conclusion about expected performance in the construct domain.
Within the current vision on assessment, the traditional concept of reliability is replaced by the notion of generalizability. Measuring reliability then becomes a question of accuracy of generalization or transfer of assessment.
The goal of the goal of using generalizability theory in reliability inquiry is to explain the consistency / inconsistency notion in scoring and to focus on understanding and reducing possible sources of measurement error.

$$\sigma^2 X_{ptr} = \sigma^2 p + \sigma^2 t + \sigma^2 r + \sigma^2 pt + \sigma^2 tr + \sigma^2 pr + ptr \text{ (totale variance)}$$

$p = person\ (student)\ ; t = task;\ r = rater$
Generalizability coefficient (reliability coefficient) : Ratio: universe score - variance/ observed score variance

Consistency of student performance on tasks?
Estimate the size of the variance components due to the interaction between students and tasks
(σ^2 St; % of total variance)
Other methods:
- Consider student scores on different tasks as items in a test and use cronbach alpha to measure the degree of consistency in scoring among tasks
- Correlate task rating and the overall mean ratings of students

Consistency of raters in their scoring?
Estimate the size of the variance component due to rating
(σ^2 r; % of total variance)
- **of students**: σ^2 pr (student*rater interaction)
- **on tasks**: σ^2 tr (task*rater interaction)
Other methods:
- Mean interrater correlations among ratings by judges who rated the same students
- Rater- total correlations (using raters as items and ratees as cases)
To explain the variability in rating, see table 2 (a framework for analyzing the variability in rating)

Procedure	4. CONSEQUENCES OF TEST USE	Techniques
Searching for evidence :	Investigating if the actual consequences are also the expected consequences	• observation of instructional strategies
What does the assessment claim to do?	How do students prepare themselves for education ?	• comparison of the quality of learning results, obtained with former test methods with learning results, using this new form of assessment.
What are the effects on the system of using the assessment other than what the assessment claims?	What kind of learning strategy is used by students?	• presenting statements of expected (and unexpected) consequences of assessment to the student population
	Which kind of knowledge is measured?	• holding semi-structured key group interviews
	Stimulates assessment the development of various skills?	
	Stimulates assessment students to apply their knowledge in realistic situations?	
	Are long term effects perceived?	
	Is breath and depth in learning actively reward, instead of mere chance?	
	Is independence stimulated by making expectations and criteria explicit?	
	Is relevant feedback provided for progress?	
	Promises for new forms of assessment	
	• encourages high quality learning, active participation	
	• encourage instructional strategies and techniques that foster reasoning, problem-solving, and communication	
	• no detriment effect on instruction because it evaluates the cognitive skill of interest directly (directness of assessment)	
	• feedback opportunities	
	• formulating clear criteria improves performance of students(transparency of assessment)	
	• encourages a sense of ownership, personal responsibility, independency and motivation	
	• ameliorates the learning climate	
	(Dochy, Segers & Slysmans, 1999; Marcoulides & Simkin, 1991; Sambell e.a., 1997; Riley, 1995 ; Topping, 1998; ....)	

Overall purpose: verify that the inferences made from the assessment are sound.
Findings should be reported in the form of an *evaluative argument* that integrates the evidence for the construct validity of the assessment
1. Is the assessment any good as measure of the construct it is interpreted to assess?
2. Should the assessment be used for the proposed use?

2.2 Validity of Assessment Scoring

The following aspect that needs to be investigated is, whether the scores are valid. In this respect, the criterion fairness plays an important role. This requires on the one hand that the assessment criteria do fit and are used appropriately: they are an adequate reflection of the criteria used by experts and the weight that is given for assessing competence. On the other hand it requires that students need to have a fair chance to demonstrate their real ability.

Potential problems are: First of all, relevant assessment criteria can be lacking. As a consequence, certain aspects of the competence at stake do not get the appropriate attention. Secondly, irrelevant (mostly personally preferred) assessment criteria can be used. Characteristic for the assessment culture is that competencies of students are assessed at different moments, in different ways, with different modes of assessment and by different assessors. In this case, potential bias in judgement will be counterweighted by the various interventions. As a result, the totality of assessment will give a more complete picture of the real competence of a person than it is the case with a single objective test, where the decision of competence is mostly reduced to a single judgement on a single moment.

2.3 Generalizability of Assessment

This step in the validating process investigates to which extent assessment can be generalised to other tasks that measure the same construct. This indicates that score interpretation is reliable and supplies evidence that assessment really measures the purport construct.

Problems that can occur are construct-underrepresentation and construct-irrelevant variance.

Construct underrepresentation means that the assessment is too small, through which important construct dimensions cannot be measured. In case of construct-irrelevant variance, the assessment is probably too broad and contains systematic variance that is irrelevant for measuring the construct (Dochy & Moerkerke, 1997). Consequently, one can discuss how broad the construct or the purport competence needs to be defined, before a given interpretation is reliable and valid.

Messick arguments that the validated interpretation gives meaning to the measure in the particular instance, and evidence on the generality of interpretation over time and across groups and settings shows how stable and thus reliable that meaning is likely to be. Other authors go much further than Messick (1989). Frederickson & Collins (1989), for example, have moved away from the idea that assessment can only be called reliable, if the interpretation can be generalised to a broader domain. They use another model where the fairness of the scoring is crucial for reliability, but where the replicability and generalizability of the performance are not.

In any case, it can be argued that an assessment using a number of authentic, representative tasks, to measure a specific competence, is less sensitive to the above mentioned problems. After all, the purported construct is directly measured. Authentic means that tasks are realistic, real life tasks. For solving these problems presented in the assessment tasks the, often complex cognitive activities experts show, are required from students.

2.4   Consequences of Assessment

Research into student learning in higher education over a number of years has provided considerable evidence to suggest that student behaviour and student learning are very much influenced by assessment (Ramsden, 1992; Marton, Hounsell & Entwistle, N., 1996; Entwistle, 1988; Biggs, 1998; Prosser & Trigwell, 1999; Scouller, 1998; Thomas & Bain, 1984). This influence of assessment can occur on different levels and depends on the function of the assessment (summative versus formative). Consequential validity, as this aspect of validity is called, addresses this issue. It implies investigation of whether the actual consequences of assessment are also the expected consequences. This can be made clear and can be brought to surface by presenting statements of expected (and unexpected) consequences of assessment to the student population, by holding semi-structured key group interviews, by recording student time logging (logging the time dedicated to assessment) or by administering self-review checklists (Gibbs, 2002). Using such methods, unexpected effects may also arise.

The influence of formative assessment (integrated within the learning process) is mainly due to the activity of looking back after the completion of the assessment task (referred to as "post-assessment-effects"). Feedback is the most important cause for these post-assessment-effects. Teachers give students information about the quality of their performance and support students in reflecting on the learning outcomes and the learning processes they are based on. When students have the necessary metacognitive knowledge and skills, teacher feedback can be reduced. Students may become capable enough to draw conclusions themselves about the quality of their learning behaviour (self-generating feedback or internal feedback), after, or even during the completion of the assessment task.

The influence of summative assessment is less obvious, but significant. Mostly, post-assessment-effects of summative assessment are small. The influence of summative assessment on learning behaviour is mainly pro-active, since students tend to adjust their learning behaviour to what they expect to be assessed. These effects can be described as pre-assessment effects, since such effects occur before assessment takes place.

An important difference between the pre- and post assessment effects is that the latter are intentional, whereas the first are rather a kind of side-effects, there the main purpose of summative assessment is not in the first place supporting and sustaining learning (but rather selection and certification of students). However, both are important effects, which need attention from teachers and instructional designers as part of the evaluation of the consequential validity of an assessment.

Nevo (1995) and Struyf, Vandenberghe and Lens (2001) point to a third kind of learning effect from assessment. Students also learn during assessment itself, because they often need to reorganise their acquired knowledge, use it in different ways to tackle new problems and to think about relations between ideas they did not discover yet during studying. When assessment stimulates them towards thinking processes of a higher cognitive level, it is possible, the authors mention, that assessment itself becomes a rich learning experience for students. This of course applies to formative, as well as to summative assessment. We can call this the true (pure) assessment-effect. Though, the true assessment effect is somewhat of a different kind than the two other effects, in that it can provide for learning but does not necessarily have a direct effect on learning behaviour, unless under the form of self-feedback as discussed earlier.

3. CHARACTERISTICS OF NEW ASSESSMENT FORMS AND THEIR ROLE IN THE CONSEQUENTIAL VALIDITY

3.1 Consequential Validity and Constructivism

Current perspectives on learning are largely influenced by constructivism. The assessment approach that is aligned with constructivist-based learning environments is sometimes referred to as assessment culture (Wolf, Bixby, Glenn, & Gardner, 1991; Kleinasser, Horsch, & Tustad, 1993; Birenbaum & Dochy, 1996). Central aspect of this assessment culture is the perception of assessment as a tool for learning. Assessment is supposed to support students in active construction of knowledge in context-rich environments, in using knowledge to analyse and solve authentic problems, in reflection. Learning so defined, is facilitated when students are participating in the process of learning and assessment as self-regulated, self-responsible learners. Finally, learning is conceived as influenced by motivation, affect and cognitive styles.

The interest for the consequential validity of assessment is in alignment with the view of assessment as a tool for learning. Evaluating the consequences of assessment is largely influenced by the characteristics of the assessment culture as part of the constructivist-based learning and teaching approach. This means that the way the consequences of assessment are defined is determined by the conceived characteristics of learning.

In the context of a constructivist-based learning environment, this leads to questions for evaluating the consequences of assessment such as: what do students understand as requirements for the assessment; how do students prepare themselves for learning and for the assessment; what kind of learning strategy is used by students; is the assessment related to authentic contexts; does the assessment stimulate students to apply their knowledge in realistic situations; does the assessment stimulate the development of various skills; are long term effects perceived; is effort, instead of mere chance actively rewarded; is breath and depth in learning rewarded; is independence stimulated by making expectations and criteria explicit; is relevant feedback provided for progress; are competencies measured, rather than memorising facts.

In the next section, the unique characteristics of new modes of assessment will be related to their effects on students' learning.

3.2 Consequential Validity and New Modes of Assessment

New modes of assessment have a positive influence on the learning of students, on the one hand by stimulating the desired cognitive skills and on the other hand by creating an environment, which has a positive influence on the motivation of students. In the following part, for each of these characteristics, we will try to unravel how they interact with personality features in order to achieve a deep and self-regulated learning behaviour.

A first characteristic of assessment is the kind of tasks that are used. New modes of assessment focus in the first place on assessing students' competencies, such as their ability to use their knowledge in a creative way to solve problems (Dochy, 1999). Tasks that are appropriate within new modes of assessment can be described as cognitive complex tasks in comparison with traditional tests items. Furthermore, assessment tasks are characterised as being real problems or authentic representations of problems in reality, whereby different solutions can be correct.

It is assumed that the different assessment demands in new modes of assessment will have a different influence on the cognitive strategies used by students. This influence of assessment on learning, called "pre-assessment-effects" earlier, is indicated by different authors with different terms ("systematic validity", "the backwash effect" and "the forward function").

The cognitive complexity of assessment tasks can have a positive influence on the students' cognitive strategies, via the pre-assessment-effect (assuming students hold the appropriate perceptions), or the post-assessment-effects (assuming there is proper feedback). There are indications that students will apply deeper learning strategies to prepare for a complex case-based exam, than for a reproductive multiple-choice exam. Students will, for example, look up for more additional information, question the content more critically and structure it more personally (McDowell, 1995; Ramsden, 1988, 1992; Sambell, McDowell, & Brown, 1997; Thomas & Bain, 1984; Trigwell & Prosser, 1991; Scouller & Prosser, 1994). The effects of the assessment demands on students' learning are mediated by the students' perceptions of these assessment demands. Research shows that there are differences between students in their capability clearly identify the nature and substance of assessment demands. Some are very adept and energetic in figuring out optimum strategies for obtaining high marks economically (students with an achieving approach), while others are less active, but take carefully note of any cues that come their way and a minority are cue deaf (Miller & Parlett, 1974). Entwistle (2000a, b) uses the terms "strategic" and "apathic" approach to indicate this difference in identification capacity of assessment demands. He poses that sensitivity to the context is required to make the best use of the opportunities for learning and to interpret the often-implicit demands by assessment tasks. To streamline the correct perception of assessment demands and the appropriate learning behaviour appears to be critical for the learning result. Nevertheless, assessment can also fulfil a supportive role. The transparency of assessment that is especially directed at clarifying the assessment expectations towards students, is one of the basic features of new assessment forms, which will be discussed further.

However, even when students correctly identify the assessment demands, they may not always be capable of adapting to it. Several studies (Martin & Ramsden, 1987; Marton & Säljö, 1976; Ramsden, 1984; Van Rossum & Schenk, 1984) have shown that students who generally use surface approaches have great difficulty adapting to assessment requirements that favour deep approaches.

Another feature of assessment tasks, namely their authentic character especially contributes to motivating students through the fact that students experience the task as more interesting and meaningful, because they realise the relevancy and usefulness of it (task-value of assessment) (Dochy & Moerkerke, 1997). Moreover, the use of authentic assessment tasks also creates the context to learn real and transferable problem solving skills, to practice and to evaluate, since these skills require a delicate interaction of different partial skills. Aiming for this interaction of partial skills, by means of "well-structured", predictable and artificial exercises, is ineffective. The authenticity of the tasks can thus be considered as an imperative condition to achieve the expert level of problem solving.

A second characteristic of new assessment forms is their formative function. The term "formative assessment" is interpreted here as encompassing all those activities explicitly undertaken by teachers, and/or students, to provide feedback to students to modify their learning behaviour in which they are engaged (see Black & Dylan, 1998). Students can also obtain feedback during instruction by actively looking for the demands of the assessment tasks. This effect is called "backwash feedback" from assessment (Biggs, 1998). This kind of feedback is not what we mean here with the term "formative assessment". The term "formative assessment" is only used for assessment that is directed at giving information to students with and after completing an assignment, and that is explicitly directed at supporting, guiding and monitoring their learning process.

The integration of assessment and learning ensures that students are encouraged to study in a more profound way during the course, at a moment that there is no time pressure, instead of "quickly learning by heart". (Askham, 1997; Dochy & Moerkerke, 1997; Sambell et al., 1997; Thomson & Falchikov, 1998). It has the advantage that students, via external and internal regulation, can get confirmation or corrective input concerning deep learning approach (Dochy & Moerkerke, 1997). External regulation refers to the assistance of the teacher by giving explicit feedback about their learning process and results. Internal regulation of the learning process is stimulated when students, based on the received feedback, reflect on the level of competency reached and on how they can improve their learning behaviour (Askham, 1997).

Moreover, feedback can also have a positive influence on the intrinsic motivation of students. The key factor to obtain these positive effects of feedback seems to be whether students perceive the primary goal of the assessment to be controlling their behaviour or providing informative and helpful feedback on their progress in learning (Deci, 1975; Keller, 1983; Ryan, Connell, & Deci, 1985). Askham (1997) points out that it is an oversimplification to think that formative assessment always leads to deep level learning and summative assessment to superficial learning. Like other authors, he argues that, in order for feedback from assessment to lead to a deep learning approach, assessment needs to be embedded in a constructivist or powerful learning environment. Birenbaum and Dochy (1996) argue that powerful learning environments are characterised by a good balance between discovery learning and personal exploration on the one hand, and systematic instruction and guidance on the other, always taking into account the individual differences abilities, needs, and motivation among students. By giving descriptive feedback- not just a grade- and organising different types of follow-up activities, a teacher creates a powerful learning environment.

A final crucial aspect of the positive influence of feedback is the way it is presented to students. Crooks (1998) identifies the following conditions for feedback in order to be effective. "First of all, feedback is most effective if it focuses on students' attention to their progress in mastering educational tasks" (p. 468-469). Therefore, it is necessary that an absolute or self-referenced norm is used, so that students can compare actual and reference levels of performance and use the feedback information to alter the gap. As has been indicated, this is also an essential condition to offer students with a normative concept of ability a possibility to realise constructive learning behaviour, since this context does not generate competitive feelings among them (which make they use defensive learning strategies). "Secondly, feedback should be given while it is still clearly relevant. This usually implies that it should be provided soon after a task is completed and that the student should then be given opportunities to demonstrate learning from feedback. Thirdly, feedback should be specific and related to its needs".

In short, formative assessment will have a positive influence on the intrinsic motivation of students and accelerate and sustain the required (or desired) constructive learning processes when it is embedded in a powerful learning environment and takes into account some crucial conditions for feedback to be effective.

A third important characteristic of assessment is the transparency of the assessment process and student involvement in the assessment process. Different authors point out that the transparency of the assessment criteria has a positive influence on the students' learning process. Indeed, "meeting criteria improves learning": if students know exactly which criteria will be used when assessing a performance, their performance will improve because they know which goals have to be attained (Dochy, 1999). As has been previously indicated, making the assessment expectations transparent towards students also has a supportive role in the correct interpretation of assessment demands that appears to be critical for the learning result (Entwistle, 2000a, b).

An effective way to make assessment transparent to students is to involve or engage them in the process of formulating criteria. As a consequence, students get a better insight in the criteria and procedures of assessment. When on top students are actually involved in the assessment process, and thus can experience practically (guided by an "expert evaluator") what it means to evaluate and judge the performance versus the criteria, this forms an additional support for the development of their self-assessment and self-regulation skills (Sadler, 1998; Gipps, 1994). Such an exercise in evaluating also contributes to the more effective use of feedback, which leads to more reflection. Elshout- Mohr (1994) arguments that students are often unwilling to give up misunderstandings, they need to be convinced through discussion, which promotes their own reflection on their thinking. If a student cannot play and carry out systematic remedial learning work for himself, he or she will not be able to make use of good formative feedback. This indicates that practising the evaluation via peer- of self-assessment is a necessary condition to come to reflection and self-regulating learning behaviour. Furthermore, Como and Rohrkemper (1985) indicate that self-regulated experiences, such as self-assessment, are closely linked with intrinsic motivation, presenting evidence that self-regulated learning experiences foster intrinsic motivation, and intrinsic motivation in turn encourages students to be more independent as learners.

Also peer-assessment, when used on a formative manner whereby the mutual evaluation functions as a support of each other's learning process, can have a positive influence on the intrinsic motivation of students.

Transparency of the assessment process, by making the judgement criteria transparent or furthermore via student involvement, can eventually lead to qualitative better learning behaviour. McDowell (1995) states that: "Assessment methods which emphasise recalling facts or the repetition of procedures are likely to lead to many students to adopt a surface approach. But also creating fear, or the lack of feedback about the progress in learning, or conflicting messages about what will be rewarded, within the assessment system, are factors that bring about the use of a surface approach. On the other hand, clearly stated academic expectations and feedback to students are more likely to encourage students to adopt a deep approach to learning".

The last characteristic of new forms of assessment is the norm that is applied. In classical testing relative standard setting has been widely used, whereby the achievement of the students is interpreted in relation to his/her fellow students. This is considered as an unfair approach within the new assessment paradigm (Gipps, 1993). Students cannot verify the achievements of other students, and cannot determine their own score. Therefore, students do not have sufficient insight in their absolute achievement. Assessment should give the student information about the level the measured competence is achieved, independent of the achievement of others. Within the new assessment paradigm, there is a tendency towards an absolute and / or self-referenced norm. The absolute norm is used both for formative and summative purposes, the self-referenced is most appropriate for formative purposes. In this respect, there is a growing tendency to use standard-setting methods where students' performances are compared with levels of proficiency of the skills measured and as defined by experts (see chapter 10).

The use of this kind of norm can give rise to a larger trust of students in the result of the assessment, when at least the transparency condition is fulfilled. The lack of a comparative norm in judgement ensures also that there is less social competition among students, and thus that there are more opportunities for collaborative learning (Gipps, 1994). The emphasis on informing students about their progress in mastery, rather than on social comparison is especially crucial for the less able students, who might otherwise receive little positive feedback. Shunk (1984) also remarks that learning and evaluation arrangements should be sufficiently flexible to ensure suitably challenging tasks for the most capable students, as otherwise they would have little opportunity to build their perception of self-efficacy. When a self-referenced norm is applied, learning- and assessment tasks can be used in a flexible way. Allowing a degree of student autonomy in choice of learning activities is a key factor in fostering intrinsic motivation, that, as has been discussed above, leads to deeper and more self-regulated learning.

4. CONCLUSION

In the so-called "new assessment era", there is a drive towards using assessment as a tool for learning. The emphasis is placed on gradually integrating instruction and assessment and on involving students as active partners in the assessment process. New developments within this framework, such as the development and use of new modes of assessment, have led to a reconsideration of quality criteria for assessment. As we argued earlier (Dierick & Dochy, 2001), the traditional criteria for evaluating the quality of tests need to be expanded in the light of the characteristics of the assessment culture.

In this contribution, we have outlined four criteria that seem necessary for evaluating the quality of new modes of assessment: the validity of assessment tasks, the validity of the assessment scoring, the generalizability of the assessment, and the consequential validity. We outlined the questions that should be posed in order to test these four criteria in more detail in our "framework for collecting evidence for and examining possible threats to construct validity".

Finally, we elaborated on the issue of "consequential validity". Until now, psychometrics has been interested in the consequences of testing only to a very small extent. The most important issue was the reliability of the measurement. Nowadays, within the edumetric approach, the consequences of the assessment do play an important role in controlling the quality of the assessment. To a growing extend, research indicates that pre- post- en true assessment effects influence the learning processes of students to a large extent. If we want students to learn more and become better learners for their lifetime, the consequential validity of the assessments is a precious jewel to handle with care. Future research on the quality of new modes of assessment, addressing the consequential validity, is needed in order to justify the widespread use of new modes of assessment.

REFERENCES

Askham, P. (1997). An instrumental response to the instrumental student: assessment for learning. *Studies in Educational Evaluation, 23* (4), 299-317.

Biggs, J. (1998). Assessment and classroom learning: A role for summative assessment? *Assessment in Education: Principles, Policy & Practices, 5*, 103-110.

Birenbaum, M., & Dochy, F. (1996). *Alternative in assessment of achievements, learning processes and prior knowledge.* Boston: Kluwer Academic Publishers.

Black, P., & Dylan, W (1998). Assessment and classroom learning. *Assessment in Education: Principles, Policy & Practices, 5*, 7-74.

Como, L., & Rohrkemper, M. (1985). The intrinsic motivation to learn. In C. Ames & R. Ames (Eds.), *Research on motivation in education* (Vol, 2). The classroom milieu (pp. 53-85). New York: Academic Press.

Cronbach, L. J. (1989). Construct validation after thirty years. In R. L. Linn (Eds.), *Intelligence: Measurement, theory and public policy.* (pp. 147-171).

Crooks, T. (1998). The impact of Classroom Evaluation Practices on Students. *Review of Educational Research, 58* (4), 438-481.

Deci, E. L.(1975). *Intrinsic motivation and self-determination in human behavior.* New York: Irvington.

Dierick, S, & Dochy, F. (2001). New lines in edumetrics: new forms of assessment lead to new assessment criteria. *Studies in Educational Evaluation, 27*, 307-329.

Dochy, F. (1999). *Instructietechnologie en innovatie van probleemoplossen: over constructiegericht academisch onderwijs.* Utrecht: Lemma.

Dochy, F., Segers, M., & Sluijsmans, D. (1999). The use of self-, peer and co-assessment in higher education: a review. *Studies in Higher Education, 24* (3), 331-350.

Dochy, F., & Moerkerke, G. (1997). The present, the past and the future of achievement testing and performance assessment. *International Journal of Educational Research, 27* (5), 415 - 432.

Dochy, F., & Mc Dowell, L. (1997). Assessment as a tool for learning. *Studies in Educational evaluation, 23*, 279-298.

Entwistle, N. J. (1988). Motivational factors in student's approaches to learning. In R. R. Smeck (Ed.), *Learning strategies and learning styles. Perspectives on individual differences* (pp. 21-51). New York: Plenum Press.

Entwistle, N. J. (2000a). Approaches to studying and levels of understanding: The influences of teaching and assessment. In J.Smart (Ed.), *Higher Education: Handbook of theory and research (XV)* (pp. 156-218). New York: Agathon Press.

Entwistle, N. J. (2000b). *Constructive alignment improve the quality of learning in higher education*. Paper presented at the Dutch Educational Research Conference, University of Leiden, May 24, 2000.

Falchikov, N. (1986). Product comparisons and process benefits of collaborative peer group and self-assessments. *Assessment and Evaluation in Higher Education, 11* (2), 146-166.

Falchikov, N. (1995). Peer Feedback Marking: Developing Peer Assessment. *Innovations in Education and Training International, 32,* (2), 395-430.

Frederiksen, J. R., & Collins, A. (1989). A system approach to educational testing. *Educational researcher, 18* (9), 27-32.

Gipps, P. (1993). *Reliability, validity and manageability in large scale performance assessment*. Paper presented at the AERA Conference, April, Atlanta.

Gipps, P. (1994). *Beyond testing: towards a theory of educational assessment*. London: The Falmer Press. (p. 119).

Gibbs, G. (2002). *Evaluating the impact of formative assessment on student learning behavior*. Paper presented at the EARLI/Northumbria Assessment conference, Longhirst, UK, August 29.

Haertel, E. H. (1991). New forms of teacher assessment. *Review of research in education, 17,* 3-29.

Kane, M. (1992). An argument-based approach to validity. *Psychological Bulletin, 112,* 527-535.

Keller, J. M. (1983).Motivational design of instruction. In C. M. Reigeluth (Ed.), *Instructional design theories and models* (pp. 383-434). Hillsdale, NJ:Erdbaum.

Kleinasser, A., Horsch, E., & Tustad, S. (1993). *Walking the talk: moving from a testing culture to an assessment culture*. Paper presented at the Annual Meeting of the American Educational Research Association, Atlanta, GA, April 1993.

Linn, R. L., Baker, E., & Dunbar, S. B. (1991). Complex, performance-based assessment: Expectations and validation criteria. *Educational Researcher, 16,* 1-21.

Marton, F., Hounsell, D. J., & Entwistle, N. J.(1996). *The experience of learning*. Edinburgh: Scottish Academic Press.

Marton,F., & Säljö, R. (1976). On qualitative differences in learning. Outcomes and process. *Britisch Journal of Educational Psychology, 46,* 4-11, 115-127.

McDowell, L. (1995 of 1996). The impact of innovative assessment on student learning. *Innovations in Education and Training International, 32* (4), 302-313.

Messick, S. (1989). Meaning and values in test validation: The science and ethics of assessment. *Educational Researcher, 18* (2), 5-11.

Miller,C. M., & Parlett, M. (1974). *Up to the mark: A study of the examination game*. London: Society for Research in Higher Education.

Nevo, D. (1995). *School-based evaluation: A dialogue for school improvement*. London: Pergamon.

Prosser, M. & Trigwell, K. (1999). *Understanding learning and teaching: The experience in higher education*. Buckingham: SRHE & Open University Press

Ramsden, P. (1984). The context of learning. In F. Marton, D.Hounsell, & N. Entwistle (Eds.), *The experience of learning*. Edinburgh: Scottish Academic press.

Ramsden, P. (1992). Student learning and perceptions of the academic environment, *Higher Education, 8,* 411-428.

Ramsden, P. (1988). *Improving learning. New perspectives*. London: Kogan page.

Ryan, R. M., Connell J. P., & Deci, E. L. (1985). A motivational analysis of self-determination and self-regulation in education. In C. Ames & R. Ames (Eds.), *Research on motivation in education: Vol2. The Classroom milieu*. New York: Academic Press.

Sadler, D. R. (1998). Formative assessment: Revisiting the territory. *Assessment in Education: Principles, Policy & Practice, 5* (1), 77-85.

Sambell, K., McDowell, L., & Brown, S. (1997). But is it fair?: An exploratory study of student perceptions of the consequential validity of assessment. *Studies in Educational Evaluation, 23* (4), 349-371.

Schunk, D (1984). Self-efficacy perspective on achievement behavior. *Educational Psychologist, 19*, 48-58.

Scouller, K. (1998). The influence of assessment method on student's learning approaches: Multiple choice question examination versus assignment essay. *Higher Education, 35*, 453-472.

Scouller, K. & Prosser, M. (1994). Students' experiences in studying for multiple-choice question examinations. *Studies in Higher Education, 19*, 267-279.

Shavelson, R. (2002). *Evaluating new approaches to assessing learning.* Invited address at the EARLI/Northumbria Assessment conference, Longhirst, UK, August 28.

Struyf, E., Vandenberghe, R., & Lens, W. (2001). The evaluation practice of teachers as a learning opportunity for students. *Studies in Educational Evaluation*, 215-238.

Thomas, P., & Bain, J. (1984). Contextual dependence of learning approaches: The effects of assessments. *Human Learning, 3*, 227-240.

Thomson, K., & Falchikov, N. (1998). Full on until the sun comes out: the effects of assessment on student approaches to studying. *Assessment & Evaluation in Higher Education, 23* (4), 379- 390.

Trigwell, K., & Prosser, M. (1991).Relating approaches to study and quality of learning approaches at the course level. *British Journal of Educational Psychology, 61*, 265-275.

Van Rossum, E. J., & Schenk, S. M. (1984). The relationship between learning conception, study strategy and learning outcome. *British Journal of Educational Psychology, 54*, 73-83.

Wolf, D., Bixby, J., Glenn, J., III, & Gardner, H. (1991). To use their minds well: Investigating new forms of student assessment. *Review of Research in Education, 17*, 31-73.

Self and Peer Assessment in School and University: Reliability, Validity and Utility

Keith Topping
Department of Psychology, University of Dundee, Scotland

1. INTRODUCTION

Self-assessment and peer assessment might appear to be relatively new forms of assessment, but in fact, they have been deployed in some areas of education for many years. For example, George Jardine, professor at the University of Glasgow from 1774 to 1826, described a pedagogical plan including methods, rules and advantages of peer assessment of writing (Gaillet, 1992). By 1999, Hounsell and McCulloch noted that over a quarter of assessment initiatives in a survey of higher education (HE) institutions involved self and/or peer assessment. Substantial reviews of the research literature on self and peer assessment have also appeared (Boud, 1995; Boud & Falchikov, 1989; Brown & Dove, 1991; Dochy, Segers, & Sluijsmans, 1999; Falchikov & Boud, 1989; Falchikov & Goldfinch, 2000; Topping, 1998).

Why should teachers, teacher educators and education researchers be interested in these developments? Can they enhance quality and/or reduce costs? Do they work? Under what conditions? This chapter explores the conceptual underpinnings and empirical evidence for the reliability, validity, effects, utility, and generalizability of self and peer assessment in schools and higher education, and by implication in the workplace and lifelong learning.

All forms of assessment should be fit for their purpose, and the purpose of any assessment is a key element in determining its validity and/or reliability. The nature and purposes of assessments influence many facets of student performance, including anxiety (Wolfe & Smith, 1995), goal orientation (Dweck, 1986), and perceived controllability (Rocklin, O'Donnell, & Holst, 1995). Of course, different stakeholders might have different purposes.

Many teachers successfully involve students collaboratively in learning and thereby relinquish some control of classroom content and management. However, some teachers might be anxious about going so far as to include self or peer assessments as part of summative assessment, where consequences follow from terminal judgements of accomplishments. By contrast, formative or heuristic assessment is intended to help students plan their own learning, identify their own strengths and weaknesses, target areas for remedial action and develop meta-cognitive and other personal and professional transferable skills (Boud, 1990, 2000; Brown & Knight, 1994). Triangulating formative feedback through the inclusion of self and peer assessment might seem to incur fewer potential threats to quality.

Reviews have confirmed the utility of formative assessment (e.g., Crooks, 1988), emphasising the importance of quality as well as quantity of feedback. Black & Wiliam (1998) concluded that assessment which precisely indicated strengths and weaknesses and provided frequent constructive individualised feedback led to significant learning gains, as compared to traditional summative assessment. The active engagement of learners in the assessment process was seen as critical, and self-assessment an essential tool in self-improvement. Affective aspects, such as the motivation to respond to feedback and the belief that it made a difference, were also important.

However, the new rhetoric on assessment might not be matched by professional practice. For example, MacLellan (2001) found that while university staff declared a commitment to the formative purposes of assessment and maintained that the full range of learning was frequently assessed, they actually engaged in practices which militated against formative and authentic assessment being fully realised.

Explorations of self and peer assessment might be driven by a need to improve quality or a need to reduce costs. These two purposes are often intertwined, since a professional assessor confronted with twice as many products to assess in the same time is likely to allocate less time to each unit of assessment, with consequent implications for the reliability and validity of the professional assessment. A peer assessor with less skill at assessment but more time in which to do it might produce an equally reliable and valid assessment. Peer feedback might be available in greater volume and with greater immediacy than teacher feedback, which might compensate for any quality disadvantage.

Beyond education, self and peer assessment (or self-improvement through self-evaluation prior to peer evaluation) are increasingly found in workplace settings (e.g. Bernadin & Beatty, 1984; Farh, Cannella, & Bedeian 1991; Fedor & Bettenhausen, 1989; Joines & Sommerich, 2001), sometimes in the guise of "Total Quality Management" or "Best Value" exercises (e.g., Kaye & Dyason, 1999). The development of such skills in school and HE should thus be transferable. University academics have long been accustomed to peer assessment of submissions to journals and conferences, the reliability and validity of which has been the subject of empirical investigation (and some concern) for many years (e.g., Cicchetti, 1991). Teachers, doctors and other professionals are often assessed by peers in vivo during practice. All of us may expect to be peer assessor and peer assessee at different times and in different contexts - or as Cicchetti (1982) more colourfully phrased it in a paper on peer review: "we have met the enemy and he is us" (p. 205).

Additionally, peer assessment in particular is connected with other forms of peer assisted learning in schools and HE. Recent research has considerably clarified the many possible varieties of peer assisted learning, their relative effectiveness in a multiplicity of contexts, and the organisational parameters crucial for effectiveness (Boud, Cohen, & Sampson, 2001; Falchikov, 2001; Topping, 1996a,b; 2001a,b; Topping & Ehly, 1998).

In this chapter, self-assessment is considered first, then peer assessment. For each practice, a definition and typology of the practice is offered, followed by a brief discussion of its theoretical underpinnings. The "accuracy", reliability and validity of the practice in schools and higher education is then considered. The research findings of the effects of the practice are then reviewed in separate sections focused on schools and higher education respectively. The research literature was searched online and manually and all relevant items included in the database for this systematic review, which consequently should have no particular bias (although space constraints do not permit mention of every relevant study by name). A summary and conclusions section for each practice relates and synthesises the findings. Finally, studies directly comparing self and peer assessment are considered, followed by an overall summary and conclusion encompassing and comparing and contrasting both practices. Evidence-based guidelines for quality implementation of self and peer assessment are then given.

2. SELF ASSESSMENT

2.1 Self Assessment - Definition, Typology and Purposes

Assessment is the determination of the amount, level, value or worth of something. Self-assessment is an arrangement for learners and/or workers to consider and specify the level, value or quality of their own products or performances.

In self-assessment, the intention is usually to engage learners as active participants in their own learning and foster learner reflection on their own learning processes, styles and outcomes. Consequently, self-assessment is often seen as a continuous longitudinal process, which activates and integrates the learner's prior knowledge and reveals developmental pathways in learning. In the longer term, it might impact self-management of learning - facilitating continuous adaptation, modification and tuning of learning by the learner, rather than waiting for others to intervene. There is evidence that graduates in employment regard the ability to evaluate one's own work as a crucial transferable skill (e.g., Midgley & Petty, 1983).

There is a large commercial market in the publication of self-test materials or self-administered revision quizzes. These are often essentially rehearsal for external summative assessment, are not used under controlled or supervised conditions, do not appear to have been rigorously evaluated, seem likely to promote superficial, mechanistic and instrumental learning, and are not our focus here. However, computerised curriculum-based self assessment test programmes which give continuous rich formative feedback to learners (often termed "Learning Information Systems") have been found effective in raising student achievement in schools (e.g., Topping, 1999; Topping & Sanders, 2000).

Self-assessment operates in many different curriculum areas or subjects. The products, outputs or performances assessed can vary - writing, portfolios, oral and/or audio-visual presentations, test performances, other skilled behaviours, or combinations of these. Where skilled behaviours in professional practice are self-assessed, this might occur via retrospective recollection or by post hoc analysis of video recordings. The self-assessment can be summative (judging a final product or performance to be correct/incorrect or pass/fail, or assigning some quantitative mark or grade) and/or (more usually) formative (involving detailed qualitative assessment of better and worse aspects, with implications for making specific onward improvements). It may be absolute (referred to external objective benchmark criteria) or relative (referring to position in relation to the products or performances of the current peer group). Boud (1989) explores the issue of whether self-assessment should form part of official student gradings, controversial if the practice is assumed to be of uncertain reliability and validity, and raising concerns about issues of power and control.

2.2 Self Assessment - Theoretical Underpinnings

What does self assessment require from students in terms of cognitive, meta-cognitive and social-affective demands? Through what processes might these benefit students? Under what conditions these processes might be optimised?

Self-assessment shares some of the characteristics of peer assessment, the theoretical underpinnings of which are discussed in detail later. Any form of assessment is a cognitively complex undertaking, requiring understanding of the goals of the task(s), the criteria for success and the ability to make judgements about the relationship of the product or performance to these. The process of self-assessment incurs extra time on task and practice. It requires intelligent self-questioning - itself cognitively demanding - and is an alternative structure for engagement with learning which seems likely to promote post hoc reflection. It emphasises learner ownership and management of the learning process, and seems likely to heighten the learner's sense of personal accountability and responsibility, as well as motivation and self-efficacy (Rogers, 1983; Schunk, 1996). All of these features are likely to enhance meta-cognition. At first sight self-assessment might seem a lonelier activity than peer assessment, but it can lead to interaction, such as when discussing assessment criteria or when the learner is called upon to justify their self-assessment to a peer or professional tutor. Such onward discussions involve constructing new schemata, moderation, norm-referencing, negotiation and other social and cognitive demands related to the mindful reception and assimilation of feedback.

2.3 Self Assessment - Reliability and Validity

This section considers the degree of correspondence between student self-assessments and the assessments made of student work by external "experts" such as professional teachers. This might be termed "accuracy" of self-assessment, if one assumes that expert assessments are themselves highly reliable and valid. As this is a doubtful assumption in some contexts (see below), it is debatable whether studies of such correspondence should be considered to be studies of reliability or validity or both or neither. This confusion is reflected in the very various vocabulary used in the literature.

There is evidence that the assessment of student products by professionals is very variable (Heywood, 1988; Newstead & Dennis, 1994; Newstead, 1996; Rowntree, 1977). Inter-rater reliabilities have been shown to vary from 0.40 to 0.63 (fourth- and eighth-grade writing portfolios) (Koretz, Stecher, Klein, & Mc Caffrey, 1994), through 0.58 to 0.87 (middle and high school writing portfolios) (LeMahieu, Gitomer, & Eresh, 1995) and 0.68 to 0.73 (elementary school writing portfolios) (Supovitz, MacGowan, & Slattery, 1997), to 0.76 to 0.94 (elementary school writing portfolios) (Herman, Gearhart, & Baker, 1993), varying with the dimensions assessed and grade level. This context should condition expectations for the "reliability" and "validity" of assessments by learners, in which the developmental process is arguably more important than "accuracy". However, Longhurst and Norton (1997) showed that tutor grades for an essay correlated quite highly (0.69 - 0.88) with deep processing criteria, while the correlation between student and tutor grades was lower (0.43).

For schoolchildren, Barnett and Hixon (1997) found age and subject differences in the reliability of self-assessment in school students. Fourth graders made relatively accurate predictions in each of three subject areas. Second graders were similar except for poor predictions in mathematics. Sixth graders made good predictions in mathematics and social studies, but not in spelling. Blatchford (1997) found race and gender differences in the reliability of self assessment in school pupils aged 7-16 years. White pupils were less positive about their own attainments and about themselves at school. While black girls showed confidence in their attainments, and had the highest attainments in reading and the study of English, white girls tended to underestimate themselves and have little confidence.

In higher education, Falchikov and Boud (1989) reported a meta-analysis of self-assessment studies which compared teacher and student marks. The degree of correspondence varied widely in different studies, from a low correlation coefficient of -0.05 to a high of 0.82, with a mean of 0.39. Some studies gave inter-assessor agreement as a percentage, and this varied from 33% to 99%, with a mean of 64%. Correspondence varied with: design and implementation quality of the study (better studies showing higher correspondence), level of the course (more advanced learners showing higher correspondence), area of study (science subjects showing higher correspondence than social science), and nature of product or performance (academic products showing higher correspondence than professional practice). Self-assessments focusing on effort rather than achievement were particularly unreliable. Overall, self-assessed grades tended to be higher than staff grades. However, more advanced students tended to under-estimate themselves.

Boud and Falchikov (1989) conducted a critical analysis of the literature on student self-assessment in HE published between 1932 and 1988. The methodological quality of studies was generally poor, although later studies tended to be better. Some studies made no mention of any explicit criteria. Where there were criteria, very many different scales were used. Some studies included ratings of student effort (of very doubtful reliability). Self-assessment sometimes appeared to be construed as the learner's guess at the professional staff assessment, rather than a rationally based independent estimate. The context for the learning to be assessed was often insufficiently described. Reports of replications were rare.

There was a tendency for earlier studies to report self-assessor over-rating and later studies under-rating. Overall, more able students tended to under-rate themselves, and weaker students to over-rate themselves by a larger amount. An interesting exception (see Gaier, 1961), found that high and low ability students produced more accurate self-assessments than middle ranking students. Boud and Falchikov (1989) found that students in the later years of courses and graduates tended to generate self-assessments more akin to staff assessments than those of students early in courses. However, those longitudinal studies which allowed scrutiny of the impact of practice in self-assessment over time showed mixed results, four studies showing improvement, three studies no improvement. Studies of any gender differences were inconclusive.

More recently, Zoller and Ben-Chaim (1997) found that students overestimated not only their abilities in the subject at hand, but also their abilities in self-assessment, as compared to tutor assessments. A review of self-assessment in medical education concluded that despite the accepted theoretical value of self-assessment, the reliability of the procedure was poor (Ward, Gruppen, & Regehr, 2002). However, several later studies have shown that the ability of students to assess themselves improves in the light of feedback or with time (Birenbaum & Dochy, 1996; Griffee, 1995). Frye, Richards, Bradley and Philp (1992) found individual students had a tendency towards over- or under-estimation in prediction of examination performance that was relatively consistent, but evolved over time with experience, maturity and self-assessment practice towards decreased overestimation and increased underestimation. Ross (1998) summarised research on self assessment, meta-analysing 60 correlations reported in the second-language testing literature. Self-assessments and teacher assessments of recently instructed ESL learners' functional English skills revealed differential validities for self-assessment and teacher assessment depending on the extent of learners' experience with the self-assessed skill.

2.4 Self Assessment in Higher Education: Effects

In considering the effects of self-assessment, the question arises of "what is a good result?" A finding that learners undertaking self-assessment have better outcomes than learners who do not, other things being equal, is clearly a "good result". A finding that learners undertaking self-assessment instead of professional assessment have outcomes as good as (if not significantly better than) learners receiving professional assessment is also arguably a "good result". However, a finding that learners undertaking self-assessment in addition to professional assessment have outcomes only as good as (and not significantly better than) learners receiving only professional assessment is not a "good result".

There are relatively few empirical studies of the effects of self-assessment. Davis and Rand (1980) compared the performance of an instructor-graded and a self-graded class. Although the self-graded class over-estimated, their overall performance was the same as the instructor-graded class. This suggests that the formative effects of self-assessment are no less than those of instructor grading, with much less effort on the part of the instructor. Sobral (1997) evaluated self-assessment of elective self-directed learning tasks, finding increased levels of self-efficacy and significant relationships to measures of deep approaches to study. Academic achievement (Grade Point Average) was significantly higher for experimental students than for controls, although not all experimental students benefited.

Marienau (1999) found longitudinal perceptions among adult learners that the experience of self-assessment strengthened commitment to subsequent competent performance, enhanced higher order skills, and fostered self-direction, illustrating that effects might not necessarily be immediate. El-Koumy (2001) investigated the effects of self-assessment on the knowledge and academic thinking of 94 English as a Foreign Language (EFL) students. Students were randomly assigned to experimental and control groups. The self-assessment group was required to assess their own knowledge and thinking before and after each lecture, during a semester. Both groups were pre- and post-tested on knowledge and academic thinking. The experimental group scored higher on both, but differences did not reach statistical significance.

2.5 Self Assessment in Schools: Effects

Similar caveats about "what is a good result?" apply here. Towler and Broadfoot (1992) reviewed the use of self-assessment in the primary school. They argued that assessment should mainly be the responsibility of the learner, and that this principle could be realistically applied in education from the early years, while emphasising the need for pupil training and a whole school approach to ensure quality and consistency. Self-assessment has indeed been successfully undertaken with some rather unlikely populations in schools, including students with learning disabilities (e.g., Lee, 1999; Lloyd, 1982; Miller, 1988) and pre-school and kindergarten children (e.g., Boersma, 1995; Mills, 1994).

Lloyd (1982) compared the effects of self assessment and self-recording as interventions for increasing the on-task behaviour and academic productivity of elementary school learning disabled students aged 9-10 years. For this population, self-recording appeared a more effective procedure than self-assessment. Miller (1988) noted that learning handicapped students tend to be passive learners. For them, self assessment included "sizing up" the task before beginning, gauging their own skill level and likelihood of success before beginning, continuous self-monitoring and assessment during task performance, and consideration of the quality of the final product or performance. Self-assessment effectiveness was seen as likely to vary according to three sets of parameters: person variables (such as age, sex, developmental skills, self-esteem), task variables (such as meaningfulness, task format, level of complexity), and strategy variables (specific strategy knowledge, relational strategy knowledge, and meta-memory).

Even with pre-school children, portfolios can be used to develop the child's own self-assessment skills and give a focus to discussions between the child and salient adults. In Mills' (1994) study, portfolios were organised around four areas of development: physical, social and emotional, emergent literacy, and logico-mathematical. For each area there was a checklist, and evidence was included to back up the checklist. At points during the school year, a professional met with each parent to discuss the portfolio, results, and goals for the child. The portfolio was subsequently made available to the child's kindergarten teacher.

Boersma (1995) described curricular modifications designed to increase students' ability to self-assess and set goals in grades K-5. Problems with self-evaluation and goal setting were documented through parent, teacher, and student surveys. Interventions included the development of a portfolio system of assessment and the implementation of reflective logs and response journals. These were successful in improving student self-evaluation and goal setting across the grades, but improvement was more marked for the older students.

Rudd and Gunstone (1993) studied the development of self-assessment skills in science and technology in third grade children. Self-assessment was scaffolded through questionnaires, concept maps and graphs created by students. Specific self-assessment concepts and techniques were introduced to the students during each term, over one academic year. Student awareness and use of skills in these classes were substantially enhanced. The teacher's role changed from controller to delegator as students became more proficient at self-assessment.

There is some evidence that engagement in self-assessment has positive effects on achievement in schools. Sink, Barnett and Hixon (1991) found that planning and self-assessment predicted higher academic achievement in middle school students. Fontana and Fernandes (1994) tested the effects of the regular use of self-assessment techniques on mathematical performance with children in 25 primary school classes. Children (n=354) in these classes showed significant improvements in scores on a mathematics test, compared to a control group (n=313). In a replication, Fernandes and Fontana (1996) found children trained in self-assessment showed significantly less dependence upon external sources of control and upon luck as explanations for school academic events, when compared to a matched control group. In addition, the experimental children showed significant improvements in mathematics scores relative to the control group.

Ninness, Ninness, Sherman and Schotta (1998) and Ninness, Ellis and Ninness (1999) trained school students in self-assessment by computer-interactive tutorials. Students received computer-displayed accuracy feedback plus reinforcement for correct self-assessments of their math performance. After withdrawal of reinforcement, self-assessment alone was found motivational, facilitating high rates and long durations of math performance. McDonald (2002) gave experimental high school students extensive training in self-assessment and using a post-test only design compared their subsequent public examination performance to that of controls, finding the self-assessment group superior.

Additionally, self-assessment in schools is not confined to academic progress. Wassef, Mason, Collins, O'Boyle and Ingham (1996) evaluated a self-assessment questionnaire for high school students on emotional distress and behavioural problems, and found it reliable in relation to staff perceptions.

2.6 Summary and Conclusions on Self Assessment

Self-assessment is increasingly widely operated in schools and HE, including with very young children and those with special educational needs or learning disabilities. It is widely assumed to enhance meta-cognition and self directed learning, but this is unlikely to be automatic. The solid evidence for this is small, although encouraging. It suggests self-assessment can result in gains in learner management of learning, self-efficacy, deep rather than superficial learning, and on traditional summative tests. Effects have been found to be at least as good as those from instructor assessment, and often better. However, effects might not be immediate and might be cumulative.

The reliability and validity of instructor assessment is not high, but that of self-assessment tends to be a little lower and more variable, with a tendency to over-estimation. The reliability and validity of self-assessment tends to be higher in relation to the ability of the learner, the amount of scaffolding, practice and feedback and the degree of advancement in the course, rather than chronological age. Other variables affecting reliability and validity include: the nature of the subject area, the nature of the product or performance assessed, the nature and clarity of the assessment criteria, the nature of assessment instrumentation, and cultural and gender differences.

In all sectors, much further development is needed, with improved implementation and evaluation quality and fuller and more detailed reporting of studies. Exploration of the effects of self-assessment is particularly needed.

3. PEER ASSESSMENT

3.1 Peer Assessment: Definition, Typology & Purposes

Assessment is the determination of the amount, level, value or worth of something. Peer assessment is an arrangement for learners and/or workers to consider and specify the level, value or quality of a product or performance of other equal-status learners and/or workers.

Peer assessment activities can vary in a number of ways, operating in different curriculum areas or subjects. The product or output to be assessed can vary - writing, portfolios, oral presentations, test performance, or other skilled behaviours. The peer assessment can be summative or formative. The participant constellation can vary - the assessors may be individuals or pairs or groups; the assessed may be individuals or pairs or groups. Directionality can vary - peer assessment can be one-way, reciprocal or mutual. Assessors and assessed may come from the same or different year of study, and be of the same or different ability. Place and time can vary - peer assessment can be formal and in class, or occur informally out of class. The objectives for the exercise may vary - the teacher may target cognitive or meta-cognitive gains, time saving, or other goals.

3.2 Peer Assessment - Theoretical Underpinnings

What does peer assessment require from students in terms of cognitive, meta-cognitive and social-affective demands? Through what processes might these benefit students? Under what conditions might these processes be optimised?

3.2.1 Feedback

The conditions under which feedback in learning is effective are complex (Bangert-Drowns, Kulik, Kulik, & Morgan, 1991; Butler & Winne, 1995; Kulhavy & Stock, 1989). Feedback can reduce errors and have positive effects on learning when it is received thoughtfully and positively. It is also essential to the development and execution of self-regulatory skills (Bangert-Drowns, et al., 1991; Paris & Newman, 1990; Paris & Paris, 2001). Butler and Winne (1995) argue that feedback serves several functions: to confirm existing information, add new information, identify errors, correct errors, improve conditional application of information, and aid the wider restructuring of theoretical schemata. Students react differently to feedback from peers and from adults (Cole, 1991; Dweck & Bush, 1976; Henry, 1979). Gender differences in responsiveness to peer feedback have also been found (Dweck & Bush, 1976), but this interacts with age (Henry, 1979).

3.2.2 Cognitive Demands

Providing effective feedback or assessment is a cognitively complex task requiring understanding of the goals of the task and the criteria for success, and the ability to make judgements about the relationship of the product or performance to these. Webb (1989) and Webb and Farivar (1994) identified conditions for effective helping: relevance to the goals and beliefs of the learner, relevance to the particular misunderstandings of the learner, appropriate level of elaboration, timeliness, comprehension by the help-seeker, opportunity to act on help given, motivation to act, and constructive activity.

Cognitively, peer assessment might create effects by gains in a number of variables pertaining to cognitive challenge and development, for assessors, assessees, or both (Topping & Ehly, 1998, 2001). These could include levels of time on task, engagement, and practice, coupled with a greater sense of accountability and responsibility. Formative peer assessment is likely to involve intelligent questioning, coupled with increased self-disclosure and thereby assessment of understanding. Peer assessment could enable earlier error and misconception identification and analysis. This could lead to the identification of knowledge gaps, and engineering their closure through explaining, simplification, clarification, summarising and cognitive restructuring. Feedback (corrective, confirmatory, or suggestive) could be more immediate, timely, and individualised. This might increase reflection and generalisation to new situations, promoting self-assessment and greater meta-cognitive self- awareness. Cognitive and meta-cognitive benefits might accrue before, during or after the peer assessment. Falchikov (1995, 2001) noted that "sleeper" (delayed) effects are possible.

3.2.3 Social Demands

Any group can suffer from negative social processes, such as social loafing, free rider effects, diffusion of responsibility, and interaction disabilities (Cohen, 1982; Salomon & Globerson, 1989). Social processes might influence and contaminate the reliability and validity of peer assessments (Byard, 1989; Falchikov, 1995; Pond, Ul-Haq, & Wade, 1995). Falchikov (2001) explores questions of role ambiguity, dissonance and conflict in relation to authority and status issues and attribution theory. Peer assessments might be partly determined by: friendship bonds, enmity or other power processes, group popularity levels of individuals, perception of criticism as socially uncomfortable or even socially rejecting and inviting reciprocation, or collusion leading to lack of differentiation. The social influences might be particularly strong with "high stakes" assessment, for which peer assessments might drift toward leniency (Farh, et al., 1991). Magin (2001a) noted that concerns about peer assessment are often focused upon the potential for bias emanating from social considerations - so-called "reciprocity effects". However, in his own study he found such effects accounted for only 1% of the variance. In any case, all these social factors require professional teacher scrutiny and monitoring. However, peer assessment demands social and communication skills, negotiation and diplomacy (Riley, 1995), and can develop teamwork skills. Learning how to give and accept criticism, justify one's own position and reject suggestions are all useful transferable social and assertion skills (Marcoulides & Simkin, 1991).

3.2.4 Affect

Both assessors and assessees might experience initial anxiety about the process. However, peer assessment involves students directly in learning, and might promote a sense of ownership, personal responsibility and motivation. Giving positive feedback first might reduce assessee anxiety and improve acceptance of negative feedback. Peer assessment might also increase variety and interest, activity and inter-activity, identification and bonding, self-confidence, and empathy with others - for assessors, assessees, or both.

3.2.5 Systemic Benefits

Peer assessment offers triangulation and per se seems likely to improve the overall reliability and validity of assessment. It can also give students greater insight into institutional assessment processes (Fry, 1990), perhaps developing greater tolerance of inevitable difficulties of discrimination at the margin. It has been contended that peer assessment is not costly in teacher time. However, other authors (e.g., Falchikov, 2001) caution that there might be no saving of time in the short to medium term, since establishing good quality peer assessment requires time for organisation, training and monitoring. If the peer assessment is to be supplementary rather than substitutional, then no saving is possible, and extra costs or opportunity costs will be incurred. However, there might be meta-cognitive benefits for staff as well as students. Peer assessment can lead staff to scrutinise and clarify assessment objectives and purposes, criteria and marking scales.

3.3 Peer Assessment - Reliability and Validity

This section considers the degree of correspondence between student peer assessments and the assessments made of student work by external "experts" such as professional teachers. Caveats regarding the use of the terms "accuracy", "reliability" and "validity" are as for self-assessment. Many purported studies of "reliability" might be considered studies of "accuracy" or "validity", comparing peer assessments with assessments made by professionals, rather than with those of other peers, or the same peers over time.

Additionally, many studies compare marks, scores and grades awarded by peers and staff, rather than upon more open-ended formative feedback. This raises concerns about the uncertain psychometric properties of such scoring scales (such as sensitivity and scalar properties), alignment of the mode of assessment with teaching and learning outcomes (i.e. relevance of the assessment), and consequently validity in any wider sense. By contrast, the reliability and validity of detailed formative feedback was explored by Falchikov (1995) and Topping, Smith, Swanson, & Elliot (2000), for example.

Research findings on the reliability and validity of peer assessment mostly emanate from studies in HE. In a wide variety of subject areas and years of study, the products and performances assessed have included:

essays (Catterall, 1995; Haaga, 1993; Marcoulides & Simkin, 1991, 1995; Orpen, 1982; Pond, et al., 1995), hypermedia creations (Rushton, Ramsey, & Rada, 1993), oral presentations (Freeman, 1995; Hughes & Large, 1993a,b; Magin & Helmore, 2001), multiple choice test questions (Catterall, 1995), practical reports (Hughes, 1995), individual contributions to a group project (Mathews, 1994; Mockford, 1994) and professional skills (Korman & Stubblefield, 1971; Ramsey, Carline, Blank, & Wenrich, 1996). Methods for computerising peer assessment are now appearing (e.g., Davies, 2000).

Over 70% of the HE studies find "reliability" and "validity" adequate, while a minority find these variable (Falchikov & Goldfinch, 2001; Topping, 1998). MacKenzie (2000) reported satisfactory reliability for peer assessment of performance in viva examinations. Magin & Helmore (2001) found inter-rater reliability for tutors making parallel assessments of oral presentations higher than that for peer assessments, but the reliability for tutors was not high (0.40 to 0.53). Magin & Helmore (2001) concluded that the reliability of summative assessments of oral presentations could be improved by combining teacher marks with the averaged marks obtained from multiple peer ratings. A tendency for peer marks to bunch around the median is sometimes noted (e.g., Catterall, 1995; Taylor, 1995). Student acceptance (or belief in reliability) varies from high (Falchikov, 1995; Fry, 1990; Haaga, 1993) to low (Rushton, et al., 1993), quite independently of actual reliability.

Contradictory findings might be explained in part by differences in contexts, the level of the course, the product or performance being evaluated, the contingencies associated with those outcomes, clarity of judgement criteria, and the training and support provided. Reliability tends to be higher in advanced courses; lower for assessment of professional practice than for academic products. Discussion, negotiation and joint construction of assessment criteria with learners is likely to deepen understanding, give a greater sense of ownership, and increase reliability (see the review by Falchikov & Goldfinch, 2000 - although Orsmond, Merry and Reiling, 2000, found otherwise). Reliability for an aggregated global peer mark might be satisfactory, but not for separate detailed components (e.g., Lejk & Wyvill, 2001; Magin, 2001b; Mockford, 1994). Peer assessments are generally less reliable when unsupported by training, checklists, exemplification, teacher assistance and monitoring (Lawrence, 1996; Pond, et al., 1995; Stefani, 1992, 1994). Segers and Dochy (2001) found peer marks correlated well with both tutor marks and final examination scores.

Findings from HE settings might not apply in other contexts. However, a number of other studies in the school setting have found encouraging consistency between peer and teacher assessments (Karegianes, Pascarella,

&, Pflaum, 1980; Lagana, 1972; MacArthur, Schwartz, & Graham, 1991; Pierson, 1967; Weeks & White, 1982).

3.4 Peer Assessment in Schools: Effects

Similar caveats about "what is a good result?" apply to peer assessment as to self-assessment. In schools, much peer assessment has focused on written products or multimedia work portfolios. A review has been provided by O'Donnell and Topping (1998).

3.4.1 Peer Assessment of Writing

Peer assessment of writing might involve giving general feedback, or going beyond that to very specific feedback about possible improvements. Peer assessment can focus on the whole written product, or components of the writing process, such as planning, drafting or editing. Studies in schools have shown less interest in reliability and validity than in HE, and more interest in effects on subsequent learner performance. Peer assessment of writing is also used with classes studying English as a Second or Additional Language (ESL, EAL) and foreign languages (Byrd, 1994; Samway, 1993).

Bouton and Tutty (1975) reported a study of the effects of peer assessment of writing with high school students. The experimental group did better than control group in a number of areas. Karegianes, et al. (1980) examined the effects of peer editing on the writing proficiency of 49 low-achieving tenth grade students. The peer edit group had significantly higher writing proficiency than students whose essays were edited by teachers. Weeks and White (1982) compared groups of grade 4 and 6 students in peer editing and teacher editing conditions. Differences were not significant, but the peer assessment group showed more improvement in mechanics and in the overall fluency of writing.

Raphael (1986) compared peer editing and teacher instruction with fifth and sixth grade students and their teachers, finding similar improvements in composition ability. Califano (1987) made a similar comparison in two fifth and two sixth grade classes, with similar results in writing ability and attitudes toward writing. Cover's (1987) study of peer editing with seventh graders found a statistically significant improvement in editing skills and attitudes toward writing. Wade (1988) combined peer feedback with peer tutoring for sixth-grade students. After training, the children could provide reliable and correct feedback, and results clearly demonstrated improvements in student writing.

Holley (1990) found peer editing of grammatical errors with grade 12 high school students in Alabama resulted in a reduction in such errors and greater student interest and awareness. MacArthur and his colleagues (1991) used peer editing in grades 4-6 in special education classrooms, which proved more effective than only regular teacher instruction. Stoddard and MacArthur (1993) demonstrated the effectiveness of peer editing with seventh and eighth grade students with learning disabilities. The quality of writing increased substantially from pre-test to post-test, and the gains were maintained at follow-up and generalised to other written work.

3.4.2 Peer Response Groups

Peer response groups are a group medium for peer assessment and feedback, obviously involving different social demands to peer assessment between paired individuals. Gere and Abbot (1985) analysed the quality of talk in response groups. Students did stay on task and provide content-related feedback. Younger students spent more time on content than did older students, who attended more to the form of the writing. However, when Freedman (1992) analysed response groups in two ninth grade classrooms, she concluded that students suppressed negative assessments of their peers.

The effects of revision instruction and peer response groups on the writing of 93 sixth grade students were compared by Olson (1986, 1990). Students receiving instruction that included both teacher revision and peer assessment wrote rough and final drafts which were significantly superior to those of students who received teacher revision only, while peer assessment only students wrote final drafts significantly superior to revision instruction only students. Rijlaarsdam (1987) and Rijlaarsdam and Schoonen (1988) made similar comparisons with 561 ninth grade students in eight different schools. Teacher instruction and peer assessment proved equally effective.

Weaver (1995) surveyed over 500 instructors about peer response groups in writing. Regardless of the stage in the writing process (early vs. late), instructors generally found peer responses to be more effective than the teacher's. In contrast, students stated they found the teacher's responses to be more helpful in all stages of writing. Nevertheless, when students could seek peer responses at the Writing Centre but not in class, their writing improved.

3.4.3 Portfolio Peer Assessment

A portfolio is "a purposeful collection of student work that exhibits the student's efforts, progress, or achievement in one or more areas. The collection must include student participation in selecting contents, the criteria for judging merit, and evidence of the student's self reflection" (Paulson, Paulson, & Meyer, 1991, p. 60). Thus, a student must be able to judge the quality of his or her own work and develop criteria for what should be included in order to develop an effective portfolio. However, there is as yet little adequate empirical literature on the effects of peer assessment of portfolios in schools.

3.4.4 Other Kinds of Peer Assessment in Schools

McCurdy and Shapiro (1992) deployed peers to undertake curriculum-based measurement in reading among 48 elementary students with learning disabilities, comparing with teacher and self-assessment. It was found that students in the self and peer conditions could collect reliable data on the number of correct words per minute. No significant differences were found between conditions. Salend, Whittaker and Reeder (1993) examined the efficacy of a consensus based group evaluation system with students with disabilities. The system involved: (a) dividing the groups into teams; (b) having each team agree on a common rating for the group's behaviour during a specified time period; (c) comparing each team's rating to the teacher's rating; and (d) delivering reinforcement to each team based on the group's behaviour and the team's accuracy in rating the group's behaviour. Results indicated that the system was an effective strategy for modifying behaviour. Ross (1995) had grade 7 students assess audio tape recordings of their own math co-operative learning groups at work. Increases in the frequency and quality of help seeking and help giving and improved students' attitudes about asking for help resulted.

3.5 Peer Assessment in Higher Education: Effects

Similar caveats about "what is a good result?" apply to peer assessment in HE as to self-assessment. In this section, studies of quantitative peer assessment are considered first, then other studies are grouped according to the product or performance assessed.

3.5.1 Peer Assessment through Tests, Marks or Grades

Hendrickson, Brady and Algozzine (1987) compared individually administered and peer mediated tests, finding scores significantly higher under the peer mediated condition. The latter was preferred by students, who found it less anxiety-provoking. Ney (1989) applied peer assessment to tests and mid-term and final exams. This resulted in improved mastery of the subject matter, and better classroom attendance. Stefani (1994) had students define the marking schedule for peer assessed experimental laboratory reports, and reported learning gains from the overall process. Catteral

(1995) had multiple choice and short essay tests peer marked by 120 marketing students. Learning gains from peer assessment were reported by 88% of participants, and impact on the ability to self-assess was reported by 76%. Hughes (1995) had first year pharmacology students use a detailed model-marking schedule. Their subsequent performance in practical work increased in comparison to previous years, whose ability on entry was identical. Segers and Dochy (2001) found no evidence of any effect of peer marking on learning outcomes.

3.5.2   Peer Assessment of Writing

In a business communication class, Roberts (1985) compared peer assessment in groups of five with staff assessment. Pre- and post-tests showed a statistically significant difference in favour of the peer condition. Falchikov (1986) involved 48 biological science students in discussion and development of essay assessment criteria. They felt the peer assessment process was difficult and challenging, but helped develop critical thinking. A majority reported increased learning and better self-organisation, while noting that it was time-consuming. The effects of teacher feedback, peer feedback and self-assessment were compared by Birkeland (1986) with 76 technicians, but no significant differences were found between conditions on test gains in paragraph writing ability. Richer (1992) compared the effects of peer group discussion of essays with teacher discussion and feedback. Grading of 174 pre- and post-test essays from 87 first year students indicated greater gains in writing proficiency in the peer feedback group. Hughes (1995) compared teacher, peer and self-assessment of written recording of pharmacology practical work, finding them equally effective.

Graner (1985) compared the effect of peer assessment and feedback in small groups to that of assessment of another's work alone using an editorial checklist. Both groups then rewrote their essays, and final grading was by staff. Both groups significantly improved from initial to final draft, and no significant difference was found between the groups. This suggests that practising critical evaluation can have generalised effects on the evaluator's own work, even in the absence of any external feedback about their own work. Chaudron (1983) compared the effectiveness of teacher feedback with feedback from peers with either English or another language as their first language. Students in all conditions showed a similar pattern of improvement. Working with 81 college students of ESL in Thailand and Hawaii, Jacobs and Zhang (1989) compared teacher, peer and self-assessment of essays. The type of assessment did not affect informational or rhetorical accuracy, but teacher and peer feedback was found to be more effective for grammatical accuracy.

3.5.3 Peer Assessment of Oral & Presentation Skills

Heun (1969) compared the effect on student self-concept of peer and staff assessment of four public speeches given by students. Compared to a control group, peer influence on the self-concept of students reached a significant level for the final speech, while instructor influence was non-significant across all four speeches. Mitchell and Bakewell (1995) found peer review of oral presentation skills led to significantly improved performance. Williams (1995) used peer assessment of oral presentations of critical incident analysis in undergraduate clinical practice nursing. Participants felt learning was enhanced, and the experience relevant to peer appraisal skills in the future working setting.

3.5.4 Peer Assessment of Group Work & Projects

Peer assessment has been used to help with the differentiation of individual contributions to small group projects (Conway, Kember, Sivan, & Wu, 1993; Falchikov, 1993; Goldfinch, 1994; Mathews, 1994), but empirical research on effects is hard to find. In a study of psychology students (Falchikov, 1993), group members and the lecturer negotiated self and peer assessment checklists of group process behaviours. Task-oriented behaviours proved easier to rate reliably than pro-social group maintenance behaviours such as facilitating the inclusion of quieter group members. Abson (1994) had marketing research students working in self-selected tutor-less groups use a simple five point rating scale on four criteria (co-operation, ideas, effort, reliability). A case study of one group suggested peer assessment might have made students work harder. Strachan and Wilcox (1996) used peer and self-assessment of group work to cope with increased enrolment in a third-year course in microclimatology. Students found this fair, valuable, enjoyable, and helpful in developing transferable skills in research, collaboration and communication.

3.5.5 Peer Assessment of Professional Skills

Peer assessment of professional skills can take place within the institution and/or out on practical placements or internships. In the latter case it is an interesting parallel to "peer appraisal" between staff in the workplace. It has been used by medical schools (e.g., Arnold, Willoughby, Calkins, Gammon, & Eberhart, 1981; Burnett & Cavaye, 1980; McAuley & Henderson, 1984), in pre-service teacher training (e.g., Litwack, 1974; Reich, 1975), and for other professions. It has also been used in short practical laboratory sessions (e.g., Stefani, 1992). Application is also reported in more exotic areas, such as applied brass jury performances (Bergee, 1993), and a range of other musical performance arts (Hunter & Russ, 1995). Lennon (1995) considered tutor, peer and self-assessments of the performance of second year physiotherapy students in practical simulations. Students rated the learning experience highly overall. Also in physiotherapy, Orr (1995) used peer assessment in role-play simulation triads. Participants rated the exercise positively, but felt some anxiety about it. Ramsey and colleagues (1996) studied peer assessment of professional performance for 187 medical interns. The process was acceptable to the subjects, and reliability adequate despite the use of self-chosen raters.

Franklin (1981) compared self, peer and expert observational assessment of teaching sessions with pre-service secondary science teachers. There were no differences between the groups in skill acquisition. A similar study by Turner (1981) yielded similar results. Yates (1982) used reciprocal paired peer feedback with fourteen special education student teachers, followed by self-monitoring. The focus was the acquisition and maintenance of the skill of giving specific praise to learning-disabled pupils. Peer feedback was effective in increasing student teachers' use of motivational praise, but not content-based praise. With self-monitoring rates of both kinds of praise were maintained. Lasater (1994) paired twelve student teachers to give feedback to each other during twelve lessons in a 5-week practicum placement. The participants reported the personal benefits to be improved self-confidence and reduced stress. The benefits to their teaching included creative brainstorming and fine-tuning of lessons, resulting in improved organisation, preparation, and delivery of lessons.

3.5.6 Computer-Assisted Peer Assessment

Wider availability of word processing and electronic mail has created opportunities for formative peer assessment in electronic draft prior to final submission, as well as distributed collaborative writing. For example, Downing and Brown (1997) describe the collaborative creation of hypertexts by psychology students, which were published in draft on the World Wide Web and peer reviewed via email. Rushton, Ramsey and Rada (1993) and Rada, Acquah, Baker, & Ramsey (1993) reported peer assessment in a collaborative hypermedia environment. Good correspondence with staff assessment was evident, but the majority of computer science students were sceptical and preferred teacher-based assessment. Brock (1993) compared feedback from computerised text analysis programmes and from peer assessment and tutoring for 48 ESL student writers in Hong Kong. Both groups showed significant growth in writing performance. However, peer interaction was rated higher for helpfulness in improving content, and peer supported students wrote significantly more words in post-intervention essays.

3.6 Summary and Conclusions on Peer Assessment

The reliability and validity of teacher assessment is not high. That of peer assessment tends to be at least as high, and often higher. Reliability tends to be higher in relation to: the degree of advancement in the course, the nature of the product or performance assessed, the extent to which criteria have been discussed and negotiated, the nature of assessment instrumentation, the extent to which an aggregate judgement rather than detailed components are compared, the amount of scaffolding, practice, feedback and monitoring, and the contingencies associated with the assessment outcome. Irrespective of relatively high reliability, student acceptance is variable.

In schools, research on peer assessment has focused less on reliability and more on effects. Students as young as grade 4 (9 years old) and those with special educational needs or learning disabilities have been successfully involved. The evidence on the effectiveness of peer assessment in writing is substantial, particularly in the context of peer editing. Here, peer assessment seems to be at least as effective in formative terms as teacher assessment, and sometimes more effective. The research on peer assessment of other learning outputs in school is as yet sparse, but merits exploration. In higher education, there is some evidence of impact of peer assessment on learning, especially in writing, sometimes greater than that of teacher assessment. In other areas, such as oral presentations, group skills, and professional skills, evidence for effects on learning are more dependent on softer data such as student subjective perceptions. Computer assisted peer assessment shows considerable promise.

In all sectors, much further development and evaluation is needed, with improved methodological quality and fuller and more detailed reporting of studies.

4. SELF VS. PEER ASSESSMENT

In Higher Education, Burke (1969) found self-assessments unreliable and peer assessments more reliable. By contrast, Falchikov (1986) found self-assessments were more reliable than peer assessments. However, Stefani (1994) found peer assessment more reliable. Saavedra and Kwun (1993) found outstanding students were the most discriminating peer assessors, but their self-assessments were not particularly reliable. Shore, Shore and Thornton (1992) found construct and predictive validity stronger for peer

Self and Peer Assessment 77

than for self-evaluations, and stronger for more easily observable dimensions than for those requiring inferential judgement. Furnham and Stringfield (1994) reported greater reliability in peer assessments by subordinates and superiors than in self-assessments. Wright (1995) found self-assessment generally yielded lower marks than peer assessment, but less so in a structured module than in a more open ended one. Lennon (1995) found a high correlation between peer assessments of a piece of work (0.85), but lesser correlations between self and peer assessment (0.61 - 0.64). However, correlations between tutor and self-assessment were even lower (0.21), and those between tutor and peer assessment modest (0.34 - 0.55). Self-assessment was associated with under-marking and bunching at the median.

In general, peer assessment seems likely to correlate more highly with professional assessment than does self-assessment, and self and peer assessments do not always correlate well. Of course, triangulation between highly correlated measures is in any event redundant, and the processes here are at least as important as the actual judgement.

5. SUMMARY AND CONCLUSION RE SELF ASSESSMENT AND PEER ASSESSMENT

Both self and peer assessment have been successfully deployed in elementary and high schools and in higher education, including with very young students and those with special educational needs or learning disabilities. The reliability and validity of assessments by professional teachers is often low. The reliability and validity of self-assessment tends to be a little lower and more variable, while the reliability and validity of peer assessment tends to be as high or higher. Self-assessment is often assumed to have meta-cognitive benefits. There is some hard evidence that it can result in improvements in the effectiveness and quality of learning, which are at least as good as gains from teacher assessment, especially in relation to writing. However, this evidence is still quite limited. There is more substantial hard evidence that peer assessment can result in improvements in the effectiveness and quality of learning, which is at least as good as gains from teacher assessment, especially in relation to writing. In other areas the evidence is softer. Of course, self and peer assessment are not dichotomous alternatives - one can lead to and inform the other. Both can offer valuable triangulation in the assessment process and both can have measurable formative effects on learning, given good quality implementation. Both need training and practice, arguably on neutral products or performances, before full implementation, which should feature monitoring and moderation. Much further development is needed, with improved implementation and evaluation quality. The following evidence-based guidelines for implementation are drawn from the research literature reviewed.

Good quality of organisation is important for implementation integrity and consistent and productive outcomes. Important planning issues evident in the literature are:
1. Clarify Purpose, Rationale, Expectations and Acceptability with all Stakeholders
2. Involve Participants in Developing and Clarifying Assessment Criteria
3. Match Participants & Arrange Contact (PA only)
4. Provide Quality Training, Examples and Practice
5. Provide Guidelines, Checklists or other tangible Scaffolding
6. Specify Activities and Timescale
7. Monitor the Process and Coach
8. Compare Aggregated Ratings, not multiple components (PA only)
9. Moderate Reliability and Validity of Judgements
10. Evaluate and Give Feedback

REFERENCES

Abson, D. (1994). The effects of peer evaluation on the behaviour of undergraduate students working in tutorless groups. In H. C. Foot, C. J. Howe, A. Anderson, A. K. Tolmie, & D. A. Warden (Eds.), *Group and Interactive Learning* (1st ed., Vol. 1, pp. 153-158). Southampton & Boston: Computational Mechanics.

Arnold, L., Willoughby, L., Calkins, V., Gammon, L., & Eberhart, G. (1981). Use of peer evaluation in the assessment of medical students. *Journal of Medical Education, 56*, 35-42.

Bangert-Drowns, R. L., Kulik, J. A., Kulik, C. C., & Morgan, M. (1991). The instructional effects of feedback in test-like events. *Review of Educational Research, 61*, 213-238.

Barnett, J. E., & Hixon, J. E. (1997). Effects of grade level and subject on student test score predictions. *Journal of Educational Research, 90* (3), 170-174.

Bergee, M. J. (1993). A comparison of faculty, peer, and self-evaluation of applied brass jury performances. *Journal of Research in Music Education, 41*, 19-27.

Bernadin, H. J., & Beatty, R. W. (1984). *Performance appraisal: Assessing human behavior at work.* Belmont, CA: Wadsworth.

Birenbaum, M., & Dochy, F. (Eds.). (1996). *Alternatives in assessment of achievement, learning processes and prior knowledge.* Boston: Kluwer Academic.

Birkeland, T. S. (1986). Variations of feedback related to teaching paragraph structure to technicians. *Dissertation Abstracts International, 47*, 4362.

Black, P., & Wiliam, D. (1998). Assessment and classroom learning. *Assessment in Education, 5* (1), 7-74.

Blatchford, P. (1997). Pupils' self assessments of academic attainment at 7, 11 and 16 years: Effects of sex and ethnic group. *British Journal of Educational Psychology, 67*, 169-184.

Boersma, G. (1995). *Improving student self-evaluation through authentic assessment.* ERIC Document Reproduction Service No. ED 393 885.

Boud, D. (1989). The role of self-assessment in student grading. *Assessment and Evaluation in Higher Education, 14* (1), 20-30.

Boud, D. (1990). Assessment and the promotion of academic values. *Studies in Higher Education, 15* (1), 101-111.

Boud, D. (Ed.). (1995). *Enhancing learning through self-assessment*. London & Philadelphia: Kogan Page.

Boud, D. (2000). Sustainable assessment: Rethinking assessment for the learning society, *Studies in Continuing Education, 22* (2), 151-167.

Boud, D., Cohen, R., & Sampson, J. (Eds.). (2001). *Peer learning in Higher Education: Learning from and with each other*. London & Philadelphia: Kogan Page.

Boud, D., & Falchikov, N. (1989). Quantitative studies of student self-assessment in higher education: A critical analysis of findings. *Higher Education, 18* (5), 529-49.

Bouton, K., & Tutty, G. (1975). The effect of peer-evaluated student compositions on writing improvement. *The English Record, 3*, 64-69.

Brock, M. N. (1993). A comparative study of computerized text analysis and peer tutoring as revision aids for ESL writers. *Dissertation Abstracts International, 54*, 912.

Brown, S., & Dove, P. (1991). *Self and Peer Assessment*. Birmingham: Standing Conference on Educational Development (SCED).

Brown, S., & Knight, P. (1994). *Assessing Learners in Higher Education*. London: Kogan Page.

Burnett, W., & Cavaye, G. (1980). Peer assessment by fifth year students of surgery. *Assessment in Higher Education, 5* (3), 273-287.

Butler, D. L., & Winne, P. H. (1995). Feedback and self-regulated learning: A theoretical synthesis. *Review of Educational Research, 65*, 245-281.

Burke, R. J. (1969). Some preliminary data on the use of self-evaluations and peer ratings in assigning university course grades. *Journal of Educational Research, 62* (10), 444-448.

Byard, V. (1989). *Power Play: The Use and Abuse of Power Relationships in Peer Critiquing*. Paper presented to Annual meeting of the conference on college composition and communication, Seattle WA, March 16-18 1989.

Byrd, D. R. (1994). Peer editing: Common concerns and applications in the foreign language classroom. *Unterrichtspraxis, 27* (1), 119.

Califano, L. Z. (1987). Teacher and peer editing: Their effects on students' writing as measured by t-unit length, holistic scoring, and the attitudes of fifth and sixth grade students. *Dissertation Abstracts International. 49* (10), 2924.

Catterall, M. (1995). Peer learning research in marketing. In S. Griffiths, K. Houston, & A. Lazenblatt (Eds.), *Enhancing student learning through peer tutoring in higher education: Section 3 - Implementing*. (1st ed., Vol. 1, pp. 54-62). Coleraine, NI: University of Ulster.

Chaudron, C. (1983). *Evaluating Writing: Effects of Feedback on Revision*. Paper presented at the Annual TESOL Convention (17th, Toronto, Ontario, March 16-19, 1983).

Cicchetti, D. V. (1982). On peer review - We have met the enemy and he is us. *Behavioral and Brain Sciences, 5* (2), 205-206.

Cicchetti, D. V. (1991). The reliability of peer-review for manuscript and grant submissions - A cross-disciplinary investigation. *Behavioral and Brain Sciences, 14* (1), 119-134.

Cohen, E. G. (1982). Expectation states and interracial interaction in school settings. *Annual Review of Sociology, 8*, 209-235.

Cole, D. A. (1991). Change in self-perceived competence as a function of peer and teacher evaluation. *Developmental Psychology, 27* (4), 682-688.

Conway, R., Kember, D., Sivan, A., & Wu, M. (1993). Peer assessment of an individual's contribution to a group project. *Assessment and Evaluation in Higher Education, 18* (1), 45-56.

Cover, B. T. L. (1987). Blue-pencil workers: the effects of a peer editing technique on students' editing skills and attitudes toward writing at the seventh grade level. *Dissertation Abstracts International, 48* (08), 1968.

Crooks, T. J. (1988). The impact of classroom evaluation practices on students. *Review of Educational Research, 58* (4), 438-481.

Davies, P. (2000). Computerized peer assessment. *Innovations in Education and Teaching International, 37* (4), 346-355.

Davis, J. K., & Rand, D. C. (1980). Self-grading versus instructor grading. *Journal of Educational Research, 73* (4), 207-11.

Dochy, F., Segers, M., & Sluijsmans, D. (1999). The use of self-, peer- and co-assessment in higher education: A review. *Studies in Higher Education, 24 (3),* 31-350.

Downing, T., & Brown, I. (1997). Learning by cooperative publishing on the World Wide Web. *Active Learning, 7,* 14-16.

Dweck, C. S. (1986). Motivational processes affecting learning. *American Psychologist, 41,* 1040-1047.

Dweck, C. S., & Bush, E. S. (1976). Sex differences in learned helplessness: Differential debilitation with peer and adult evaluators. *Developmental Psychology, 12* (2), 1.

El-Koumy, A. S. A. (2001). *Effects of Student Self-Assessment on Knowledge Achievement and Academic Thinking.* Paper presented at the Annual Meeting of the Integrated English Language Program-II (3rd, Cairo, Egypt, April 18-19, 2001). ERIC Document Reproduction Service No. ED 452 731.

Falchikov, N. (1986). Product comparisons and process benefits of collaborative peer group and self assessments. *Assessment and Evaluation in Higher Education, 11* (2), 146-166.

Falchikov, N. (1993). Group process analysis - Self and peer assessment of working together in a group. *Educational & Training Technology International, 30 (3),* 275-284.

Falchikov, N. (1995). Peer feedback marking - Developing peer assessment. *Innovations in Education and Training International, 32,* 175-187.

Falchikov, N. (2001). *Learning together: Peer tutoring in higher education.* London & New York: RoutledgeFalmer.

Falchikov, N., & Boud, D. (1989). Student self-assessment in higher education: A meta-analysis. *Review of Educational Research, 59* (4), 395-430.

Falchikov, N., & Goldfinch, J. (2000). Student peer assessment in Higher Education: A meta-analysis comparing peer and teacher marks. *Review of Educational Research, 70* (3), 287-322.

Farh, J., Cannella, A. A., & Bedeian, A. G. (1991). Peer ratings: The impact of purpose on rating quality and user acceptance. *Group and Organization Studies, 16,* 367-386.

Fedor, D. B., & Bettenhausen, K. L. (1989). The impact of purpose, participant preconceptions, and rating level on the acceptance of peer evaluations. *Group and Organization Studies, 14,* 182-197.

Fernandes, M., & Fontana, D. (1996). Changes in control beliefs in Portuguese primary school pupils as a consequence of the employment of self-assessment strategies. *British Journal of Educational Psychology, 66,* 301-313.

Fontana, D., & Fernandes, M. (1994). Improvements in mathematics performance as a consequence of self-assessment in Portuguese primary-school pupils. *British Journal of Educational Psychology, 64,* 407-417.

Franklin, C. A. (1981). Instructor versus peer feedback in microteaching on the acquisition of confrontation; illustrating, analogies, and use of examples; and question-asking teaching skills for pre-service science teachers. *Dissertation Abstracts International, 42*, 3565.

Freedman, S. W. (1992). Outside-in and inside-out: Peer response groups in two ninth grade classes. *Research in the Teaching of English, 26*, 71-107.

Freeman, M. (1995). Peer assessment by groups of group work. *Assessment and Evaluation in Higher Education, 20*, 289-300.

Fry, S. A. (1990). Implementation and evaluation of peer marking in higher education. *Assessment and Evaluation in Higher Education, 15*, 177-189.

Frye, A. W., Richards, B. F., Bradley, E. W., & Philp, J. R. (1992). The consistency of students self-assessments in short-essay subject-matter examinations. *Medical Education, 26* (4), 310-316.

Furnham, A., & Stringfield, P. (1994). Congruence of self and subordinate ratings of managerial practices as a correlate of supervisor evaluation. *Journal of Occupational and Organizational Psychology, 67*, 57-67.

Gaier, E. L. (1961). Student self assessment of final course grades. *Journal of Genetic Psychology, 98* (1), 63-67.

Gaillet, L. I. (1992). *A foreshadowing of modern theories and practices of collaborative learning: The work of the Scottish rhetorician George Jardine.* Paper presented at the 43rd Annual Meeting of the Conference on College Composition and Communication, Cincinnati OH, March 19-21 1992.

Gere, A. R., & Abbot, R. D. (1985). Talking about writing: The language of writing groups. *Research in the Teaching of English, 19*, 362-379.

Goldfinch, J. (1994). Further developments in peer assessment of group projects. *Assessment and Evaluation in Higher Education, 19* (1), 29-35.

Graner, M. H. (1985). Revision techniques: Peer editing and the revision workshop. *Dissertation Abstracts International, 47*, 109.

Griffee, D. T. (1995). *A longitudinal study of student feedback: Self-assessment, Course Evaluation and Teacher Evaluation.* Longman: Birmingham, Alabama.

Haaga, D. A. F. (1993). Peer review of term papers in graduate psychology courses. *Teaching of Psychology, 20* (1), 28- 32.

Hendrickson, J. M., Brady, M. P., & Algozzine, B. (1987). Peer-mediated testing: The effects of an alternative testing procedure in higher education. *Educational and Psychological Research, 7* (2), 91-102.

Henry, S. E. (1979). Sex and locus of control as determinants of children's responses to peer versus adult praise. *Journal of Educational Psychology, 71* (5), 605.

Herman, J., Gearhart, M., & Baker, E. (1993). Assessing writing portfolios: Issues in the validity and meaning of scores. *Educational Assessment, 1* (3), 201-224.

Heun, L. R. (1969). *Speech Rating as Self-Evaluative Behavior: Insight and the Influence of Others.* PhD Dissertation, Southern Illinois University.

Heywood, J. (1988). *Assessment in higher education.* Chichester: John Wiley.

Holley, C. A. B. (1990). The effects of peer editing as an instructional method on the writing proficiency of selected high school students in Alabama. *Dissertation Abstracts International, 51* (09), 2970.

Hounsell, D., & McCulloch, M. (1999). Assessing skills in Scottish higher education. In E. Dunne (Ed.) *The learning society: International perspectives on core skills in higher education* (pp.149-158). London: Kogan Page.

Hughes, I. E. (1995). Peer assessment. *Capability, 1* (3), 39-43.

Hughes, I. E., & Large, B. J. (1993a). Staff and peer-group assessment of oral communication skills. *Studies in Higher Education, 18*, 379-385.

Hughes, I. E., & Large, B. J. (1993b). Assessment of students' oral communication skills by staff and peer groups. *New Academic, 2* (3), 10-12.

Hunter, D., & Russ, M. (1995). Peer assessment in performance studies. In S. Griffiths, K. Houston, & A. Lazenblatt (Eds.), *Enhancing Student Learning Through Peer Tutoring in Higher Education: Section 3 - Implementing.* (1st ed., Vol. 1, pp. 63-65). Coleraine, NI: University of Ulster.

Jacobs, G., & Zhang, S. (1989). *Peer Feedback in Second Language Writing Instruction: Boon or Bane?* Paper presented at the Annual Meeting of the American Educational Research Association (San Francisco, CA, March 27-31, 1989).

Joines, S. M. B., & Sommerich, C. M. (2001). Comparison of self-assessment and partnered-assessment as cost-effective alternative methods for office workstation evaluation. *International Journal of Industrial Ergonomics, 28* (6), 327-340.

Karegianes, M. L., Pascarella, E. T., & Pflaum, S. W. (1980). The effects of peer editing on the writing proficiency of low-achieving tenth grade students. *Journal of Educational Research, 73* (4), 203-207.

Kaye, M. M., & Dyason, M. D. (1999). Achieving a competitive focus through self-assessment. *Total Quality Management, 10* (3), 373-390.

Koretz, D., Stecher, B., Klein, S., & Mc Caffrey, D. (1994). The Vermont portfolio assessment program: Findings and implications. *Educational Measurement, 13* (3), 5-16.

Korman, M., & Stubblefield, R. L. (1971). Medical school evaluation and internship performance. *Journal of Medical Education, 46*, 670-673.

Kulhavy, R. W., & Stock, W. A. (1989). Feedback in written instruction: The place of response certitude. *Educational Psychology Review, 1*, 279-308.

Lagana, J. R. (1972). *The development, implementation and evaluation of a model for teaching composition which utilizes individualized learning and peer grouping.* Unpublished doctoral thesis, University of Pittsburgh, Pittsburgh, PA.

Lasater, C. A. (1994). Observation feedback and analysis of teaching practice: Case studies of early childhood student teachers as peer tutors during a preservice teaching practicum. *Dissertation Abstracts International, 55*, 1916.

Lawrence, M. J. (1996). *The effects of providing feedback on the characteristics of student responses to a videotaped high school physics assessment.* Unpublished doctoral thesis, Rutgers University, New Brunswick, NJ.

Lee, B. (1999). *Self-assessment for pupils with learning difficulties.* Slough UK: National Foundation for Educational Research.

Lejk, M., & Wyvill, M. (2001). Peer assessment of contributions to a group project: A comparison of holistic and category-based approaches. *Assessment & Evaluation in Higher Education, 26* (1), 61-72.

LeMahieu, P., Gitomer, D. H., & Eresh, J. T. (1995). Portfolios in large-scale assessment: Difficult but not impossible. *Educational Measurement, 14* (3), 11-16, 25-28.

Lennon, S. (1995). Correlations between tutor, peer and self assessments of second year physiotherapy students in movement studies. In S. Griffiths, K. Houston, & A. Lazenblatt (Eds.), *Enhancing Student Learning Through Peer Tutoring in Higher Education: Section 3 - Implementing.* (1st ed., Vol. 1, pp. 66-71). Coleraine, NI: University of Ulster.

Litwack, M. (1974). A study of the effects of authority feedback, peer feedback, and self feedback on learning selected indirect-influence teaching skills. *Dissertation Abstracts International, 35*, 5762.

Lloyd, J. (1982). Reactive effects of self-assessment and self-recording on attention to task and academic productivity. *Learning Disability Quarterly, 5* (3), 216-27.
Longhurst, N., & Norton, L. S. (1997). Self-assessment in coursework essays. *Studies in Educational Evaluation, 23* (4), 319-330.
MacArthur, C. A., Schwartz, S. S., & Graham, S. (1991). Effects of a reciprocal peer revision strategy in special education classrooms. *Learning Disabilities Research and Practice, 6* (4), 201-210.
MacKenzie, L. (2000). Occupational therapy students as peer assessors in viva examinations. *Assessment & Evaluation in Higher Education, 25* (2), 135- 147.
MacLellan, E. (2001). Assessment for learning: the differing perceptions of tutors and students. *Assessment & Evaluation in Higher Education, 26* (4), 307-318.
Magin, D. J. (2001a). Reciprocity as a source of bias in multiple peer assessment of group work. *Studies in Higher Education, 26* (1), 53-63.
Magin, D. J. (2001b). A novel technique for comparing the reliability of multiple peer assessments with that of single teacher assessments of group process work. *Assessment and Evaluation in Higher Education, 26* (2), 139-152.
Magin, D., & Helmore, P. (2001). Peer and teacher assessments of oral presentation skills: How reliable are they? *Studies In Higher Education, 26* (3), 287-298.
Marcoulides, G. A., & Simkin, M. G. (1991). Evaluating student papers: The case for peer review. *Journal of Education for Business, 67*, 80-83.
Marcoulides, G. A., & Simkin, M. G. (1995). The consistency of peer review in student writing projects. *Journal of Education for Business, 70* (4), 220-223.
Marienau, C. (1999). Self-assessment at work: Outcomes of adult learners' reflections on practice. *Adult Education Quarterly, 49* (3), 135-146.
Mathews, B. P. (1994). Assessing individual contributions - Experience of peer evaluation in major group projects. *British Journal of Educational Technology, 25*, 19-28.
McAuley, R. G., & Henderson, H. W. (1984). Results of the peer assessment program of the college of physicians and surgeons of Ontario. *Canadian Medical Association Journal, 131*, 557-561.
McCurdy, B. L., & Shapiro, E. S. (1992). A comparison of teacher-monitoring, peer-monitoring, and self-monitoring with curriculum-based measurement in reading among students with learning-disabilities. *Journal of Special Education, 26* (2), 162-180.
McDonald, B. (2002). *Self assessment and academic achievement*. Unpublished Ph.D. thesis, University of the West Indies, Cave Hill, Barbados, West Indies.
Midgley, D. F., & Petty, M. (1983). *Final report on the survey of graduate opinions on general education*. Kensington: University of New South Wales.
Miller, M. (1988). *Self-Assessment in students with learning handicaps*. Paper presented at the Annual Convention of the Council for Exceptional Children (66th, Washington, DC, March 28-April 1, 1988). ERIC Document Reproduction Service No. EC 210 383.
Mills, L. (1994). *Yes, It Can Work!: Portfolio Assessment with Preschoolers*. Paper presented at the Association for Childhood Education International Study Conference (New Orleans, LA, March 30-April 2, 1994). ERIC Document Reproduction Service No. ED 372 857.
Mitchell, V. W., & Bakewell, C. (1995). Learning without doing - Enhancing oral presentation skills through peer-review. *Management Learning, 26*, 353-366.
Mockford, C. D. (1994). The use of peer group review in the assessment of project work in higher education. *Mentoring and Tutoring, 2* (2), 45-52.
Newstead, S. E. (1996). The psychology of student assessment. *The Psychologist, 9*, 543-7.
Newstead, S., & Dennis, I. (1994). Examiners examined. *The Psychologist, 7*, 216-9.

Ney, J. W. (1989). *Teacher-Student Cooperative Learning in the Freshman Writing Course.* ERIC Document Reproduction Service.

Ninness, H. A. C., Ellis, J., & Ninness, S. K. (1999). Self-Assessment as a learned reinforcer during computer interactive math performance - An Experimental analysis. *Behavior Modification, 23* (3), 403-418.

Ninness, H. A. C., Ninness, S. K., Sherman, S., & Schotta, C. (1998). Argumenting computer-interactive self-assessment with and without feedback. *Psychological Record, 48* (4), 601-616.

O'Donnell, A. M., & Topping, K. J. (1998). Peers assessing peers: Possibilities and problems. In K. J. Topping, & S. W. Ehly (Eds.), *Peer assisted learning* (Chapter 14, pp. 255-278). Mahwah NJ: Lawrence Erlbaum.

Olson, V. L. B. (1986). The effects of revision instruction and peer response groups on the revision behaviors, quality of writing and attitude toward writing of sixth grade students. *Dissertation Abstracts International, 47* (12), 4310.

Olson, V. L. B. (1990). The revising processes of sixth-grade writers with and without peer feedback. *Journal of Educational Research, 84* (1), 1.

Orpen, C. (1982). Student vs lecturer assessment of learning: A research note. *Higher Education, 11,* 567-572.

Orr, A. (1995). Peer assessment in a practical component of physiotherapy education. In S. Griffiths, K. Houston, & A. Lazenblatt (Eds.), *Enhancing Student Learning Through Peer Tutoring in Higher Education: Section 3 - Implementing.* (1st ed., Vol. 1, pp. 72-78). Coleraine, NI: University of Ulster.

Orsmond, P., Merry, S., & Reiling, K. (2000). The use of student derived marking criteria in peer and self-assessment. *Assessment & Evaluation in Higher Education, 25* (1), 23-38.

Paris, S. G., & Newman, R. S. (1990). Developmental aspects of self-regulated learning. *Educational Psychologist, 25,* 87-102.

Paris, S. G., & Paris, A. H. (2001). Classroom applications of research on self-regulated learning. *Educational Psychologist, 36* (2), 89-101.

Paulson, E. L., Paulson, P. R., & Meyer, C. A. (1991). What makes a portfolio a portfolio? *Educational Leadership, 48* (5), 60-63.

Pierson, H. (1967). *Peer and teacher correction: A comparison of the effects of two methods of teaching composition in grade 9 English classes.* Unpublished doctoral thesis, New York University, New York.

Pond, K., Ul-Haq, R., & Wade, W. (1995). Peer review: A precursor to peer assessment. *Innovations in Education and Training International, 32,* 314-323.

Rada, R., Acquah, S., Baker, B., & Ramsey P. (1993). Collaborative learning and the MUCH System. *Computers and Education, 20,* 225-233.

Ramsey, P. G., Carline, J. D., Blank, L. L., & Wenrich M. D. (1996). Feasibility of hospital-based use of peer ratings to evaluate the performances of practicing physicians. *Academic Medicine, 71,* 364-370.

Raphael, T. E. (1986). *The impact of text structure instruction and social context on students' comprehension and production of expository text.* East Lansing, MI: Institute for Research on Teaching, Michigan State University.

Reich, R. (1975). The effect of peer feedback on the use of specific praise in student-teaching. *Dissertation Abstracts International, 37,* 925.

Richer, D. L. (1992). The effects of two feedback systems on first year college students' writing proficiency. *Dissertation Abstracts International, 53,* 2722.

Rijlaarsdam, G. (1987). *Effects of peer evaluation on writing performance, writing processes, and psychological variables.* Paper presented at the 38th Annual Meeting of the

Conference on College Composition and Communication, Atlanta GA, March 19-21, 1987.
Rijlaarsdam, G., & Schoonen, R. (1988). *Effects of a teaching program based on peer evaluation on written composition and some variables related to writing apprehension.* Amsterdam: Stichting Centrum voor Onderwijsonderzoek, Amsterdam University.
Riley, S. M. (1995). Peer responses in an ESL writing class: Student interaction and subsequent draft revision. *Dissertation Abstracts International, 56,* 3031.
Roberts, W. H. (1985). The effects of grammar reviews and peer-editing on selected collegiate students' ability to write business letters effectively. *Dissertation Abstracts International, 47,* 1994.
Rocklin, T. R., O'Donnell, A. M., & Holst, P. M (1995). Effects and underlying mechanisms of self-adapted testing. *Journal of Educational Psychology, 87,* 103-116.
Rogers, C. R. (1983). *Freedom to learn for the 80s.* Columbus OH: Charles E. Merrill.
Ross, J. A. (1995). Effects of feedback on student behavior in cooperative learning groups in a grade-7 math class. *Elementary School Journal, 96* (2), 125-143.
Ross, S. (1998). Self-assessment in second language testing: A meta-analysis and analysis of experiential factors. *Language Testing, 15* (1), 1-20.
Rowntree, D. (1977). *Assessing students: How shall we know them?* London: Harper & Row.
Rudd, T. J., & Gunstone, R. F. (1993). *Developing self-assessment skills in grade 3 science and technology: The importance of longitudinal studies of learning.* Paper presented at the Annual Meetings of the National Association for Research in Science Teaching (Atlanta, GA, April 15-18, 1993) and the American Educational Research Association (Atlanta, GA, April 12-16, 1993). ERIC Document Reproduction No. ED 358 103.
Rushton, C., Ramsey, P., & Rada, R. (1993). Peer assessment in a collaborative hypermedia environment - A case-study. *Journal of Computer-Based Instruction, 20,* 75-80.
Saavedra, R., & Kwun, S. K. (1993). Peer evaluation in self-managing work groups. *Journal of Applied Psychology, 78,* 450-462.
Salend, S. J., Whittaker, C. R., & Reeder, E. (1993). Group evaluation - a collaborative, peer-mediated behavior management system. *Exceptional Children, 59* (3), 203-209.
Salomon, G., & Globerson, T. (1989). When teams do not function the way they ought to. *International Journal of Educational Research, 13,* 89-99.
Samway, K. D. (1993). This is hard, isn't it - children evaluating writing. *Tesol Quarterly, 27* (2), 233-257.
Schunk, D. H. (1996). *Learning theories: An educational perspective* (2nd ed.). Englewood Cliffs NJ: Prentice-Hall.
Segers, M., & Dochy, F. (2001). New assessment forms in problem-based learning: The value-added of the students' perspective. *Studies in Higher Education, 26* (3), 327-343.
Shore, T. H., Shore, L. M., & Thornton, G. C. (1992). Construct validity of self evaluations and peer evaluations of performance dimensions in an assessment center. *Journal of Applied Psychology, 77,* 42-54.
Sink, C. A., Barnett, J. E., & Hixon, J. E. (1991). Self-regulated learning and achievement by middle-school children. *Psychological Reports, 69* (3), 979-989.
Sobral, D. T. (1997). Improving learning skills: A self-help group approach. *Higher Education, 33* (1), 39-50.
Stefani, L. A. J. (1992). Comparison of collaborative self, peer and tutor assessment in a biochemistry practical. *Biochemical Education, 20* (3), 148-151.
Stefani, L. A. J. (1994). Peer, self and tutor assessment - Relative reliabilities. *Studies in Higher Education, 19* (1), 69-75.

Stoddard, B., & MacArthur, C. A. (1993). A peer editor strategy - guiding learning-disabled students in response and revision. *Research in the Teaching of English, 27* (1), 76-103.

Strachan, I. B., & Wilcox, S. (1996). Peer and self assessment of group work: Developing an effective response to increased enrolment in a third-year course in microclimatology. *Journal of Geography in Higher Education, 20* (3), 343- 353.

Supovitz, J. A., MacGowan, A., & Slattery J. (1997). Assessing agreement: An examination of the interrater reliability of portfolio assessment in Rochester, New York. *Educational Assessment, 4* (3), 237-259.

Taylor, I. (1995). Understanding computer software: Using peer tutoring in the development of understanding of some aspects of computer software. In S. Griffiths, K. Houston, & A. Lazenblatt (Eds.), *Enhancing Student Learning Through Peer Tutoring in Higher Education: Section 3 - Implementing.* (1st ed., Vol. 1, pp. 87-89). Coleraine, NI: University of Ulster.

Topping, K. J. (1996a). The effectiveness of peer tutoring in further and higher education: A typology and review of the literature. *Higher Education, 32* (3), 321-345. (Also in S. Goodlad. (Ed.). (1998). *Mentoring and tutoring by students.* London & Stirling VA: Kogan Page.)

Topping, K. J. (1996b). *Effective peer tutoring in further and higher education* (SEDA Paper 95). Birmingham: Staff and Educational Development Association.

Topping, K. J. (1998). Peer assessment between students in college and university. *Review of Educational Research, 68* (3), 249-276.

Topping, K. J. (1999). Formative assessment of reading comprehension by computer: Advantages and disadvantages of the Accelerated Reader software. *Reading OnLine* (I.R.A.) [Online]. Available www.readingonline.org/critical/topping/ [November 4]. (hypermedia).

Topping, K. J. (2001a). *Peer assisted learning: A practical guide for teachers.* Cambridge MA: Brookline Books.

Topping, K. J. (2001b). *Tutoring by peers, family and volunteers.* Geneva: International Bureau of Education, United Nations Educational, Scientific and Cultural Organisation (UNESCO). [Online] Available: www.ibe.unesco.org/International/Publications/EducationalPractices/prachome.htm [January 1] (Also in translation in Chinese and Spanish).

Topping, K. J., & Ehly, S. W. (Eds.). (1998). *Peer-assisted learning.* Mahwah NJ & London UK: Lawrence Erlbaum.

Topping, K. J., & Ehly, S. W. (2001). Peer assisted learning: A framework for consultation. *Journal of Educational and Psychological Consultation, 12* (2), 113-132.

Topping, K. J., & Sanders, W. L. (2000). Teacher effectiveness and computer assessment of reading: Relating value added and learning information system data. *School Effectiveness and School Improvement, 11* (3), 305-337.

Topping, K. J., Smith, E. F., Swanson, I., & Elliot, A. (2000). Formative peer assessment of academic writing between postgraduate students. *Assessment and Evaluation in Higher Education, 25* (2), 149-169.

Towler, L, & Broadfoot, P. (1992). Self-assessment in the primary school. *Educational Review, 44* (2), 137-151.

Turner, R. F. (1981). The effects of feedback on the teaching performance of preservice teachers involved in a microteaching experience. *Dissertation Abstracts International, 42,* 3116.

Wade, L. K. (1988). An analysis of the effects of a peer feedback procedure on the writing behavior of sixth-grade students. *Dissertation Abstracts International, 50* (05), 2181.

Ward, M., Gruppen, L., & Regehr, G. (2002). Measuring self-assessment: Current state of the art. *Advances in Health Sciences Education, 7* (1), 63-80.

Wassef, A., Mason, G., Collins, M. L., O'Boyle, M., & Ingham, D. (1996). In search of effective programs to address students' emotional distress and behavioral problems: Student assessment of school-based support groups. *Adolescence, 31* (12), 1-16.

Weaver, M. E. (1995). Using peer response in the classroom: Students' perspectives. *Research and Teaching in Developmental Education, 12*, 31-37.

Webb, N. M. (1989). Peer interaction and learning in small groups. *International Journal of Educational Research, 13*, 13-40.

Webb, N. M., & Farivar, S. (1994). Promoting helping behavior in cooperative small groups in middle school mathematics. *American Educational Research Journal, 31*, 369-395.

Weeks, J. O., & White, M. B. (1982). *Peer editing versus teacher editing: Does it make a difference?*. Paper presented at Meeting of the North Carolina Council of the International Reading Association, Charlotte NC, March 7-9, 1982.

Williams, J. (1995). Using peer assessment to enhance professional capability. In M. Yorke (Ed.), *Assessing Capability in Degree and Diploma Programmes.* (1st ed., Vol. 1, pp. 59-67). Liverpool: Centre for Higher Education Development, Liverpool John Moores University.

Wolfe, L., & Smith, J. K. (1995). The consequence of consequence: Motivation, anxiety, and test performance. *Applied Measurement in Education, 8*, 227-242.

Wright, L. (1995). All students will take more responsibility for their own learning. In S. Griffiths, K. Houston, & A. Lazenblatt (Eds.), *Enhancing Student Learning Through Peer Tutoring in Higher Education: Section 3 - Implementing.* (1st ed., Vol. 1, pp. 90-92). Coleraine, NI: University of Ulster.

Yates, J. A. (1982). The effects of peer feedback and self-monitoring on student teachers' use of specific praise. *Dissertation Abstracts International, 43*, 2321.

Zoller, Z., & Ben-Chaim, D. (1997). *Student self-assessment in Science Examinations: Is it compatible with that of teachers?* Paper presented at the meeting of the European Association for Research on Learning and Instruction, Greece, Athens, August 26-30.

# A Framework for Project-Based Assessment in Science Education

Yehudit J. Dori
Department of Education in Technology and Science Technion, Israel institute of Technology, Haifa, Israel and Center for Educational Computing Initiatives, Massachusetts Institute of Technology, Cambridge, USA.

1. INTRODUCTION

Assessment in science education is commonly applied to evaluate students, but it can be applied also to teachers and to entire schools (Nevo, 1994; 1995). Lewy (1996) proposed that assessment be based on a set of tasks, including oral responses, writing essays, performing data manipulations with technology-enhanced equipment, and selecting a solution from a list of possible options. Similarly, in science education, student assessment is defined as a collection of information on students' outcomes, both while learning is taking place – formative assessment, and after the completion of the learning task – summative assessment (Tamir, 1998). Commenting on the common image of testing and assessment, Black (1995) has noted:

"Many politicians, and most of the general public, have a narrow view of testing and assessment. The only mode which they know and understand is the conventional test, which is seen as a reliable and cheap way of comparing schools and assessing individuals." (p. 462).

Since the mid eighties, educators' awareness of the need to modify the traditional testing system in schools has increased throughout the western world (Black, 1995, 1995a). In the mid nineties, multiple choice items and standardized test scores have been supplemented with new methods, such as portfolios, hands-on, performance assessment and self-assessment (Baxter & Shavelson, 1994; Baxter, Shavelson, Goldman, & Pine, 1992; Ruiz-Primo, & Shavelson, 1996; Tamir, 1998). Researchers are investigating the effect of alternative assessment on various groups of students (Birenbaum & Dochy, 1996; Flores & Comfort, 1997; Lawrenz, Huffman, & Welch, 2001; Shavelson & Baxter, 1992). Other studies investigate how teaching and learning in science can benefit from embedded assessment (Treagust, Jacobowitz, Gallagher, & Parker, 2001).

Focusing on assessment that is based on projects carried out by students, students' assessment is closely related to alternative assessment, as defined by Nevo (1995, p. 94):

"In alternative assessment, students are evaluated on the basis of their active performance in using knowledge in a creative way to solve worthy problems. The problems have to be real problems."

Projects are becoming an acceptable means for both teaching and assessment. Being a school-wide endeavour, a project performance can serve as a means to assess not only the individual student or student team, but also the school that designed and carried out the project. Formative and summative evaluations should provide information for project planning, improvement and accountability (Nevo, 1983, 1994).

In recent years, new modes of assessment have been receiving researchers' attention. When new modes of assessment are applied, students are evaluated on the basis of their ability to solve authentic problems. The problems have to be non-routine and multi-faceted with no obvious solutions (Nevo, 1995). In science education, the term embedded assessment is used in conjunction with alternative assessment when referring to an ongoing process that emphasizes integration of assessment into teaching. Teachers can use embedded assessment to guide instructional decisions for making adjustments to teaching plans in response to the level of students' conceptual understanding (Treagust et al., 2001). The combination of alternative and embedded assessment can potentially yield a powerful and effective set of tools for fostering higher order thinking skills (Dori, 2003). In this chapter, the term "new modes of assessment" refers to the combination of alternative and embedded assessment modes.

The amount and extent of decisions that high-level administrators and education experts make is deemed by many as too high. To counter this trend, schools and teachers should be more involved in new developments in assessment methods (Nevo, 1995). Indeed, the American National Science Education Standards (NRC, 1996) indicated that teachers are in the best position to use assessment data to improve classroom practice, plan curricula, develop self-directed learners, report students' progress, and research teaching practices. According to Treagust et al. (2001), the change from a testing culture, which is the common assessment practice, to an assessment culture, be it embedded or alternative, is a systemic change. Such a profound reform mandates that teachers, educational institutions, and testing agencies rethink the educational agenda and the role of assessment.

As participants in authentic evaluation, researchers cannot set aside their individual beliefs and viewpoints, through which they observe and analyse the data they gathered (Guba & Lincoln, 1989). To attenuate the bias such individual beliefs cause, evaluation of educational projects should include opinions of the various stakeholders as part of the data.

This chapter exposes the reader to three studies, which outline a framework for project-based assessment in science education. The studies describe new modes of assessment that integrate alternative and embedded assessment, as well as internal and external assessment. Three different types of population participated in the studies: six-graders, high school students, and junior-high school teachers in Israel. In all three studies, emphasis was placed on assessing higher order thinking skills. The studies are summarized and conclusions that enable the construction of a project-based assessment framework are drawn.

2. PROJECT-BASED ASSESSMENT

Project-based curriculum constitutes an innovative teaching/learning method, aimed at helping students cope with complex real world problems (Keiny, 1995; McDonald & Czerniac, 1994). The project-based teaching/learning method involves both theoretical and practical aspects. It can potentially convey to students explicit and meaningful subject matter content from various disciplines, in a concrete yet comprehensive fashion. Project-based learning enhances higher order thinking skills, including data analysis, problem solving, decision-making and value judgement. Blumenfeld, Marx, Soloway and Krajcik (1996) argued that project-related tasks tend to be collaborative, open-ended and to generate problems with answers that are often not predetermined. Knowledge generation is emphasized as students pose questions, gather data and information, interpret findings, and use evidence to draw conclusions. Individuals, groups, or the whole class, can actively participate in creating unique artefacts to represent their understanding of natural and scientific phenomena that the project involves.

Project-based learning is discussed in several studies (Cheung, Hattie, Bucat, & Douglas, 1996; Solomon, 1993). Through their active participation in the project execution process, students are encouraged to form original opinions and express individual standpoints. The project fosters students' awareness of system complexity, and encourages them to explore the consequences of their own values (Zoller, 1991).

While engaging in project-based curriculum, the traditional instruments for measuring literacy do not fully convey the essence of student performance. Mitchell (1992b) pointed out the contribution of authentic assessment to learning process, and the advantages of engaging students, teachers, and schools in the assessment processes. Others have proposed various means aimed at assessing project-based learning (Black, 1995; Nevo, 1994; Tal, Dori, & Lazarowitz, 2000).

3. DEVELOPING AND ASSESSING HIGHER ORDER THINKING SKILLS THROUGH CASE STUDIES

Project-based assessment is suited to foster and evaluate higher order thinking skills. Resnick (1987) stated that although it is difficult for researchers to define higher order thinking skills, these skills could be recognized when they occur. Based on Costa (1985), Dillon (1990), Shepardson and Pizzini (1991), and using TIMSS (Shorrocks-Taylor, & Jenkins, 2000) taxonomy, research projects described in this chapter involved both low- and high-level assignments. A low-level assignment is usually characterized as having a definite, clear, "correct" response, so it is relatively easy to assess and grade it, and the assessment is, for the most part, on the objective and "neutral" side. Low-level assignments require the students to recall knowledge and understand concepts. The opposite is true for high-level assignments, where the variability and range of possible and acceptable responses is far greater, as there is not just one "school solution". High-level assignments are open-ended and require various combinations of application, analysis, synthesis, inquiry, and transfer skills. Open-ended assignments promote different types of student learning and demonstrate that different types of knowledge are valued (Resnick & Resnick, 1992; Wiggins, 1989; Zohar & Dori, 2003). By nature, assessing high-level assignments is more demanding and challenging than that of low-level ones, as the assessing teachers need to be able to embrace different viewpoints and accept novel ideas or original, creative responses that they had not thought of before.

Performance assessment by means of case studies is a recommended practice in science teaching (Dori, 1994; Herried, 1994). The case study method, which fosters a constructivist learning environment, was applied in

A Framework for Project-Based Assessment in Science Education 93

all three studies described in this chapter. The underlying principles of the case study method are similar to the problem based or context based method. Starting at business and medical schools, the case study method has become a model for effective learning and gaining the attention of the student audience. Case studies are usually real stories, examples for us to study and appreciate, if not emulate. They can be close-ended, demanding correct answers, or open-ended, with multiple solutions because the data involves emotions, ethics or politics. Examples of such open-ended cases include global warming, pollution control, human cloning and mission to Mars (Herried, 1997). In addition to case studies, for assessing teachers, peer assessment was applied, while for assessing students, self-assessment was applied. Peer assessment encourages group interaction and critical review of relative performance and increases responsibility for one's own learning (Pond & Ul Haq, 1997).

In what follows, three studies on project-based assessment are described. All three studies are discussed with respect to Research goal and objectives, Research setting, Assessment, and Method of analysis and findings. Conclusions are then drawn for all three studies together, which provide the basis for the project-based assessment framework. The subjects of studies I and II are students, while those of study III are teachers. Students in study I are from elementary school, in study II – high school students, and in study III – Junior-high school science teachers. Studies I and III investigate the process of project-based learning and assessment, while the value-added of study II is the quasi experimental design with control groups.

4. STUDY I – ELEMENTARY SCHOOL INDUSTRY-ENVIRONMENT COLLABORATIVE PROJECTS

Many science projects concern the natural environment and advance knowing and appreciating nature as part of science education or education in general (Bakshi & Lazarowitz, 1982; Hofstein & Rosenfeld, 1996). Fewer sources refer to the industrial environment as part of contemporary human environment that allows project-based learning (Posch, 1993; Solomon, 1993). This study investigated six-graders who were engaged in an industry-environment project. The project involved teams of students, guided by parents, who chose, planned and manufactured industrial products, while accounting for environmental concerns. The community played a major role in influencing the theme selection, mentoring the students, and assessing their performances (Dori & Tal, 2000).

4.1 Research Goal and Objectives

The research goal was to develop and implement a project-based assessment system for interdisciplinary learning processes. The objectives were to investigate the extent to which the projects contributed to developing students' higher order thinking skills (Tal et al., 2000).

4.2 Research Setting

The study was carried out in a community elementary school, where the community and students' parents select portions from the school national curricula, develop local curricula and design enrichment materials. As Darling-Hammond (1994) noted, in schools that undergo restructuring, teachers are responsible for students' learning processes and for using authentic tools to assess how students learn and think.

The study was conducted during three years and included about 180 six-grade students. The main theme of the project focused on the nearby high-tech Industrial Park, which is located in a natural mountainous region. The industry-environment part of the school-based curriculum was taught formally during three months. The last eight weeks of that period were dedicated to the project, which was carried out informally, after school hours, in parallel with the formal learning.

The objectives of the project included:
- Exposing the students to real world problems and "learning-by-doing" activities;
- Enabling students to summarize their learning by means of a portfolio and an exhibition that is open to the community at large;
- Encouraging critical thinking and system approach; and
- Fostering collaborative learning and social interactions among students, parents and community.

Each year, parents, students and teachers chose together a new industry and environment related theme. Then, the teachers divided the students into heterogeneous teams of 10-12 students. Within each team every student was expected to help and be helped (Lazarowitz & Hertz-Lazarowitz, 1998). The student teams, guided by volunteer parents and community experts, were involved in studying the scientific background related to the project themes. Examples of project themes included building a battery manufacturing plant in the neighbourhood, designing a plant for recycled paper products and products related to road and public areas improvement. Teachers observed the group activities and advised the mentors. Experts from the community met and advised team members.

All the decisions, processes, corresponding, programs and debates were documented by the students and collected for inclusion in the project portfolio, which was presented as an important part of the exhibition. Both the portfolio and the exhibition are new modes of assessment, suggested by Nevo (1995) as a means to encourage students' participation in the assessment process, and to foster interaction between the students and their assessors. The last stage of the project was the presentation and exhibition in an "industrial exhibition". The exhibition was planned according to the ideas of Sizer (1992), who suggested that the educational exhibition serves as a means of meaningful learning and demonstrates various student skills in the cognitive, affective and communicative domains.

4.3 Assessment

The assessment system comprised a suite of several assessment tools: pre- and post-case studies; CHEAKS questionnaire; portfolio content analysis; community expert assessment; and students' self-assessment (Dori & Tal, 2000).

The knowledge part of CHEAKS – Children's Environmental Attitude and Knowledge Scale questionnaire (Leeming, Dwyer, & Bracken, 1995) was used to evaluate students' knowledge and understanding of key terms and concepts. Higher order thinking skills and learning outcomes were assessed through the use of pre- and post-case studies, in which the students were required to exercise decision making and demonstrate awareness of system complexity.

The community experts assessed the exhibition, and the teachers assessed the portfolio. The project-team score was based on the assessment of the team's portfolio and its presentation in the exhibition. The portfolios were also sent to an external national competition. Teachers assessed the case studies, while the students assessed themselves. The case study and self-assessing determined the individual student score.

We used the content analysis to analyse the case studies and team portfolios. The purpose of the content analysis was to determine the level of student knowledge and understanding of key terms and concepts, one's ability to analyse industrial-environmental problems, decision-making ability and awareness of system's complexity.

Several interviews with students, parents and teachers were conducted right after team meetings for raising specific questions or issues concerning the assessment process that needed to be discussed and clarified. This way, the process itself generated additional ideas for assessment criteria and methods.

In developing the project-based assessment system, we focused on two main aspects. One aspect was whether the assessed object is the individual student or the entire team. The other aspect was the assessing agent, i.e., who does the assessment (Tal et al., 2000). The findings and conclusions were reflected in the final design of the system, which is summarized in Table 1.

Table 1. Methods, agents and criteria of students' assessment

Assessment Method	Agent	Criteria
Pre- and post-case studies	Teachers	• Identifying the problem • Posing questions/raising hypotheses • Argumentation • Taking a stand
CHEAKS	Researchers	• Knowledge and understanding
Portfolio	Teachers and experts	• Using scientific, industrial and environmental concepts • Collecting and presenting scientific data • System thinking presented in industry-environment relationships, problem solving and decision making • Reflective thinking • Conceptualization - the contribution of the project to understanding environmental problems
Exhibition	Community experts	• Product design • Exhibition design • Manufacturing process • Marketing and advertisement • Environmental considerations • Team oral presentations
Self assessment	Students	• Attending team meetings • Listening to team members • Collaboration with peers • Initiatives within the team • The number of chapters in the portfolio to which the student contributed • Team activities to which the student contributed • Specific individual contribution • Project's influence on student's school life • Socialization difficulties within team; relation to mentors

The case study, with which students had to deal, concerned establishing a new industrial area in the Western Galilee. The Regional Council was in favor of the plan, as it would benefit the surrounding communities. Many objected to the plan. One assignment that followed this case study was as follows: Think of possible reasons for rejecting the plan and write them down.

4.4 Method of Analysis and Findings

In students' responses to the case study assignment, 21 different hypotheses were identified and classified into two categories: economic/societal and environmental. Each category was classified into one of three possible levels: high, intermediate, and low. Three science education experts validated the classification of the hypotheses by category and level. Table 2 presents examples for the hypotheses classified by the categories and levels. All four teachers who participated in the project established the scientific content validity of a random sample of 20% of the assignments.

Table 2. Hypothesis examples by category and level for the case study assignment

Category – Level	Example
Economic/societal – low	New plants might cause competition.
Economic/societal – high	It could be unprofitable and a waste of money, because the region is unpopulated and there are enough industrial areas.
Environmental – low	In the western Galilee there are beautiful landscapes. Industry might harm them.
Environmental – intermediate	Air and water pollution may damage wildlife. Microbes and diseases may affect us.
Environmental – high	Industrial areas cause a lot of damage like air and water pollution by chemicals. This is dangerous to people and wildlife. Landscapes are ruined to allow construction and the whole region changes.

Relating the case study scores to the CHEAKS scores, Figure 1 shows the pre- and post-course CHEAKS knowledge and the case study scores. The improvement for the entire population was significant ($p < 0.0001$) in both knowledge, required in the CHEAKS, and high order thinking skills, required in the case study assignments. The lack of a control group was compensated for by using additional assessment modes, including portfolio, exhibition with external reviewers, and participation in national competition.

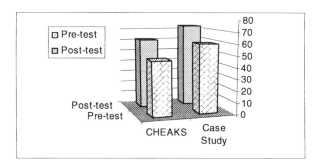

Figure 1. Pre- and post-course CHEAKS knowledge and the case study scores

The portfolios were written summaries of the teams' work, and served for teachers as assessment objects. The portfolios included description of the teamwork, product selection, surveys enacted, product design, planning the manufacturing process, planning the marketing strategy, financial programs and marketing policy. It also contained reflections about the team collaborative work and contribution to understanding of technological-environmental conflicts.

To analyse the portfolios we defined five general categories. Three of them are listed below along with examples from the portfolios.

- System thinking presented in industry-environment relationships: *"Highly developed industry improves the economical situation, which, in turn, improves life quality."* (economic/societal consideration).
- Reflective thinking: *"We learned what team work is and how hard it is. We understand what taking responsibility and independent work means."*
- Conceptualisation: *"All these problems lead us to think about solutions, because if we will do nothing, reality will be worse. Having done our project, we are more aware of the environment... There are many possible solutions: regulations of the Ministry of the Environment about air pollution and chemical waste, preventing the emission of poisonous pollutants from industrial plant chimneys by raising them and by installing filters, using environment friendly materials, monitoring clearance and treatment of sewerage and solid waste..."*

The process of analysing the portfolios according to these categories enabled us to grade the portfolios on a scale of five levels at each category. In the first year, three of the five teams got a high grade, while the other two got an intermediate grade. In the second year, two of the seven teams got a high grade, three teams got an intermediate grade, and two teams got a lower grade.

The distinction between internal assessors and external evaluators does not imply any value judgment regarding the advantage of one over the other (Nevo, 1995). Both functions of assessment were important in our study. However, the external evaluation helped us in demonstrating accountability.

External reviewers evaluated the portfolios, in addition to the school teachers. A national competition of student works was conducted by the Yubiler Institute in the Hebrew University in Jerusalem. All five portfolios of the pilot study (the first of three years) were submitted to the competition, and achieved the highest assessment of "Special Excellence". The open dialogue between the internal assessment – both formative and summative, and the external one – formative evaluation, as presented in the portfolio assessment, contributed to the generalization power of our method (Tal et al., 2000).

The project's pinnacle was the industrial exhibition, where the students presented the products and portfolios. The presentation included the manufactured products, the marketing program and tools, the students' explanations about the environmental solutions and the teams' portfolios. Various experts and scholars from the community were invited to assess the teams' work. They represented various domains, including industry, economy, education, design and art.

The teachers and guiding parents had prepared a list of assessing criteria (see Table 1). The community experts interviewed each team and suggested two best teams for each criterion and were impressed by the students' ability to acquire technological-environmental literacy.

To accomplish a comprehensive project-based assessment, we elicited the students' point of view through a self-assessment questionnaire. The questionnaire was developed as a result of negotiations with the students. The students suggested self-assessment criteria, including attendance in team meetings, listening to team members, cooperation with peers, and initiatives within the team and the sub-teams. In this questionnaire, almost all the students ranked as very high the criteria of initiatives within the sub-team (86%), attendance in team meetings (74%), and cooperation with peers (67%). They indicated that the project helped them develop reflections and self-criticism.

5. STUDY II – "MATRICULATION 2000" PROJECT IN ISRAEL: SCHOOL-BASED ASSESSMENT

Matriculation examinations in Israel have been the dominant summative assessment tool of high school graduates over the last half-century. The grades of the matriculation examinations, along with a psychometric test (analogous to SAT in the USA), are a critical factor in college and university admission requirements. This nationwide battery of tests is conducted centrally in seven or eight different courses, including mathematics, literature, history, English and at least one of the sciences (physics, chemistry, or biology). The Ministry of Education determines the goals and contents of each course. A national committee appointed by the Ministry is charged with composing the corresponding tests and setting criteria for their grading. This leaves the schools and the teachers with little freedom to modify either the subject matter or learning objectives. However, students' final grade in the matriculation transcript for each course is the average of the school grade in the course and the pertinent matriculation examination grade.

A national committee headed by Ben-Peretz (1994) examined the issue of the matriculation examinations from two aspects: Pedagogical – quality of teaching, learning and assessment; and socio-cultural – the number and distribution of students from diverse communities eligible for the Matriculation Diploma.

Addressing the socio-cultural aspect, several researchers (Gallard, Viggiano, Graham, Stewart, & Vigiliano, 1998; Sweeney & Tobin, 2000) have claimed that educational equity goes beyond the notion of equal opportunity and freedom of choice. The way learning is fostered should be examined to verify whether students are allowed to use all the intellectual tools that they bring with them to the classrooms.

The Ben-Peretz Committee indicated that in their current format, the matriculation examinations do not reflect the depth of learning that takes place in many schools, nor do they measure students' creativity. The Committee's recommendations focused, among other issues, on providing high schools with increased autonomy to apply new modes of assessment instead of the nationwide matriculation examination. The school-based assessment would combine traditional examinations with new modes of assessment in a continuous fashion throughout high school, from 10th through 12th grade. In addition to tests, the proposed assessment methods included individual projects, portfolios, inquiry laboratory experiments, assignments involving teamwork, and article analysis. The Committee called for nominating exemplary schools, which would be mentored and monitored by experts in one, two, or three courses in each school. The school grades in those courses would be recognized as the standard matriculation grades.

As a result of the Ben-Peretz Committee's recommendations, the Ministry of Education launched a five-year project, titled "Matriculation 2000." The Project aimed at developing deep understanding, higher order thinking skills, and students' engagement in learning through changes in both teaching and assessment methods. During the period of 1995-1999, 22 schools from various communities participated in the project. The courses taught in these schools under the umbrella of the "Matriculation 2000" Project were chemistry, biology, English, literature, history, social studies, bible, and Jewish heritage. In the liberal art courses the most prevalent assessment methods were individual projects, portfolios, assignments involving teamwork, and presentations to peers. In the science courses, portfolios, inquiry laboratory experiments, assignments involving teamwork, concept maps, and article analysis were the most widely used assessment methods.

An expert group accompanied each school, providing the teachers with guidance in teamwork, school-based curriculum, and new modes of assessment. These expert groups were themselves guided and managed by an overseeing committee headed by Ben-Elyahu (1995).

5.1 Research Goal and Objectives

The research goal was to investigate students' learning outcomes in chemistry, biology and literature in the "Matriculation 2000" Project. The assumption was that new modes of assessment have some effect on students' outcomes in both affective and cognitive domains. The research objectives were to investigate the attitudes that students express toward new modes of teaching and assessment applied in the Project and the Project's effect on students' achievements in chemistry, biology, and literature.

5.2 Research Setting

The research population included two groups of students in six heterogeneous high schools (labelled School A through School F) out of the 22 exemplary schools that participated in the "Matriculation 2000" Project.

The **first** group, which included students from 10th and 12th grades (N = 561) served the investigation regarding the effect of the project on students' affective domain. The Israeli high school starts at 10th grade and ends at 12th grade, therefore, tenth grade was the first year a student participated in the Project, while 12th grade was the last one. The schools represented a variety of communities, academic levels, and sectors, including urban, secular, religious, and Arab schools. The students from these six schools responded to attitude questionnaires regarding new modes of teaching and assessment applied in the Project (see Table 3). The courses taught in these six schools were chemistry, biology, literature, history, and Jewish heritage.

All the students in the Project who studied chemistry and biology took the courses at the highest level of 5 units, which is comparable to an Honors class in the US high school system and A Level in the European system. Most of the students who studied liberal arts courses took them at the basic level of 2 units, which is comparable to Curriculum II in the US high school system and O Level in the European system. In School D and School E, one science course and one liberal arts course were taught in the framework of the "Matriculation 2000" Project. In the other four schools, only one course was taught as part of the Project.

The **second** group, described in Table 3, served the investigation regarding the effect of the project on students' cognitive domain. In four out of the six experimental schools, 214 12th graders responded to achievement tests in chemistry (School A), biology (School E) and literature (School B and School C). These students served as the experimental group for assessing achievements. Another 162 12th grade students, who served as a control group, responded to identical achievement tests in chemistry, biology, and literature. These students were from two high schools (labelled G and H), which did not participate in the Project, but were at an academic and socio-economic level comparable to that of the experimental schools.

To enable comparison between the experimental and control groups, the grades that teachers had given to the students in the participating schools were collected. No significant differences in chemistry and biology between the experimental and the control students were found. In literature, there was a significant difference in favour of the experimental students ($\bar{X}_{experiment} = 77$, $\bar{X}_{control} = 72$; $t = -2.89$, $p < 0.01$), but since the difference was only 0.05 (5 points out of 100) and this difference was found only in literature, the experimental and the control groups were considered as identical.

The new modes of assessment applied in the experimental schools included portfolios, individual projects, team projects, written and oral tests, class and homework assignments, self assessments, field trips, inquiry laboratory activities, concept maps, scientific article reviews, and project presentations. These methods were integrated into the teaching throughout the school year and therefore constituted an embedded assessment. The most prevalent methods, as reported by teachers and principals, were written tests, class and homework assignments, individual or group projects, and scientific article reviews. In chemistry, the group effort was a mini-research project that spanned over half a year. Students were required to raise a research question, design an experiment to investigate the question, carry it out, and draw conclusions from its outcomes. In biology and literature, the students presented individual projects to their peers in class and expert visitors in an exhibition. In literature, the project included selecting a subject, stage and play it, or design a related visual artefact (Dori, Barnea, & Kaberman, 1999).

To gain deeper insight into the Project setting, consider School A. In the middle of 10th grade, students in this school were given the opportunity to decide whether they wanted to elect chemistry at the Honors level. Students who chose this option, studied in groups of 20 per class for eight hours per week throughout 11th and 12th grades. These students focused on 80% of the topics included in the national, standard Honors chemistry curriculum, but they were exposed also to many more laboratory activities as well as to scientific articles. New modes of assessment were embedded throughout the curriculum. The teachers' teamwork included a weekly two-hour meeting for designing the individual and group projects, their theoretical and laboratory contents, along with additional tools and criteria for assessing these projects. Teachers graded the projects and scientific article reviews according to topic rather than class affiliation. They claimed that this process increased the level of reliability and objectivity of the grades.

Table 3. Research population and research instruments.

Research group	School	Courses taught in Project	N	Attitude questionnaire administered to grade	Courses for which achievement test was administered	N	Achievement test administered to grade
experi-mentel	A	Chemistry	70	10th & 12th	Chemistry	59	12th
	B	Literature	103	10th & 12th	Literature	44	12th
	C	Literature	47	10th & 12th	Literature	30	12th
	D	Biology History	160	10th & 12th	-	-	-
	E	Biology History	121	10th & 12th	Biology	81	12th
	F	Jewish heritage	60	10th & 12th	-	-	-
	Total		561			214	
control	G	-	-	-	Chemistry Biology History	38 35 30	12th
	H	-	-	-	Biology Literature	30 29	12th
	Total					162	

5.3 Assessment

Students' attitudes toward the Project are defined in this chapter as students' perceptions of the teaching and assessment methods in the Project. These, in turn, were measured by the attitude questionnaires. Following preliminary visits to the six experimental schools, an initial open attitude questionnaire was composed and administered to 50 11th grade students in one of the experimental schools. Based on responses to this preliminary questionnaire, a comprehensive two-part questionnaire was compiled. Since there were 160 items, they were divided into two questionnaires of 80 items each, with each question in one questionnaire having a counterpart in the other. In part A, items were clustered in groups of five or six. Each group of items referred to a specific question that represented a category. Examples for such categories are *"What is the importance of the Project?"* and *"Compare the teaching and assessment methods in the Project with the traditional ones."* Part B included positive and negative items that were mixed in a random fashion throughout the questionnaire without specifying the central topic being investigated. For the purpose of analysis, negative items were reversed. All items in part B were later classified into the following categories: students' motivation and interest, learning environment, students' responsibilities and freedom of choice, and variety of teaching and assessment methods. Students were asked to rank each item in both parts on a scale of 1 to 5, where 1 was "totally disagree" and 5 was "totally agree."

The effect of the Project on students' performance in chemistry, biology, and literature was measured through a battery of achievement tests. These tests were administered to the experimental and control 12th grade students. Three science education/literature experts constructed each test and set criteria for its grading. Two other senior science/literature teachers, who were on sabbatical that year and hence did not teach any course, read and graded each test independently. The final test grade was computed as the average of the scores assigned by the two graders. In less than 5% of the cases, the difference between the grades each senior teacher assigned was greater than 10 (out of 100) points. In such cases, one of the experts, who participated in constructing the test and the criteria, also evaluated the test independently. This expert, who took in account the three grades, determined the final grade.

The assignments in these tests referred to a given unseen: case study (in science) or a poem (in literature) and were categorized into low-level and high-level ones.

5.4 Method of Analysis and Findings

The scores of students' responses to the attitude questionnaires (where the scale was between 1 and 5) ranged from 2.50 to 4.31 on the average per item. Following are several items and their corresponding scores. The item that scored the highest was "The assessment in the Project is based on a variety of methods rather than a single test". Another high-scoring item (4.15) was "The Project enables self expression through creative projects and assignments, not just tests." A relatively high score of 3.84 was obtained for the item reading "Many students take part in class discussions." The lowest score, 2.50, was obtained for the item regarding the existence of a joint teacher-student team whose task was to determine the yearly syllabi. Another item that scored low (2.52) was "Students ask to reduce the number of weekly lessons per course."

Table 4 presents students' attitude scores for the three highest scoring categories (formulated as questions) in part A. The highest four items in each category are listed in descending order, along with their corresponding scores. The average per category accounts for all the items in the category, not just the four ones that are listed. Therefore, the category average is somewhat lower than the average of the four highest items in the category.

To find out about the types of changes participating students would like to see taking place in the "Matriculation 2000" Project, two complementary questions were posed. The first question, which appeared in one of the questionnaire versions, was *"What would you like to increase or modify in the Project?"* It included the items "Include more courses", "Include more creative projects", "Include more teamwork", "Keep it as it is now", and "Discontinue it in our school". The two responses "strongly agree" and "agree" were classified as "for" and the two responses "strongly disagree" and "disagree" were classified as "against". More than 60% of the students who responded to this question preferred that the Project include more courses and more creative projects. More than half of the students disagreed or strongly disagreed with the item calling for the Project to be discontinued in their own school.

The complementary question, which appeared in the other questionnaire version, was *"What would you like to reduce or eliminate from the Project"* More than half of the students agreed with the item "Reduce time-consuming projects" while 43% agreed with the item "Eliminate all examinations". About 80% were against cancelling the Project, 57% disagreed with the item "Do not reduce anything," and 52% disagreed with reducing the amount of teamwork.

Table 4. Students' attitude scores for the three highest scoring categories

Category	Score	Highest scoring items	Item Score
Effect of Project on teaching and assessment methods	**3.95**	The assessment in the Project is based on a variety of methods rather than a single test	4.31
		The Project enables both individualized work and teamwork	3.95
		The teacher is more attentive to what students have to say	3.89
		The teacher serves as both advisor and tutor rather than just lecturer	3.70
Advantages of the new modes of teaching methods	**3.89**	There are many opportunities to improve students' grades	4.15
		There is no pressure to cover all the course material for the Matriculation examination	3.95
		Students are more active and involved in the learning process	3.94
		Students are more active and involved in the assessment process	3.83
Project importance	**3.70**	The Project enables self-expression through creative projects and assignments, not just tests	4.15
		The Project reduces stress by eliminating the course's Matriculation examination	4.11
		The Project exposes students to interesting learning approaches	3.85
		Students with learning difficulties get a chance of obtaining better scores	3.84

Overall, students were supportive of continuing the Project, were in favour of adding more courses into the Project's framework, and preferred more creative projects and fewer examinations. At the same time, students were in favour of decreasing the workload.

Table 5. Average scores of high-level assignments by Project course and research group

Course	Research Group	N	High-level assignments	
			$\bar{X}$	T
Chemistry	Experimental	59	82.7	6.14**
	Control	38	63.5	
Biology	Experimental	81	64.0	4.87**
	Control	65	55.7	
Literature	Experimental	74	61.7	3.07*
	Control	59	59.2	

*$p < 0.01$; **$p < 0.0001$

A detailed description of the various types of the assignments is provided elsewhere (Dori, 2003). The findings regarding students' achievements have shown that the experimental students achieved significantly higher score than their control group peers on assignments that required knowledge and understanding. For example, in chemistry, experimental and control students scored an average of 80.1 and 57.4, respectively ($p < 0.001$). In high-level assignments (see Table 5), the differences between the two research groups were greater and the gap was wider compared with the respective differences in knowledge-level assignments. Some of this wide gap can be attributed to the fact that a lower level of knowledge in the control group hampered their achievements at the high-level assignments. At any rate, this gap is a strong indication that the "Matriculation 2000" Project has indeed attained one of its major objectives, namely, fostering higher order thinking skills. This outcome is probably a result of the fact that students worked on the projects both individually and in teams, and had to discuss scientific issues that relate to daily life complex problems.

The national standardized system and the school-based assessment system co-exist, but for the purpose of university admission, a weighted score is computed, which accounts for both matriculation examination score (which embodies an element of school assessment) and the score of a battery of standard psychometric tests.

6. STUDY III – JUNIOR-HIGH SCHOOL SCIENCE TEACHERS PROJECTS

In response to changes in science and technology curricula, the Israeli Ministry of Education decided (Harari, 1994) to provide teachers with a series of on-going Science and Technology workshops of one day per week for a period of three academic years. This research followed two groups of teachers who participated in these workshops at the Department of Education in Technology and Science at the Technion. The workshops included three types of enrichment: theoretical, content knowledge and pedagogical content knowledge (Shulman, 1986).

6.1 Research Goal and Objectives

The goal of the research was to study various aspects of the new modes of assessment approach in the context of teachers' professional development. The research objectives were to investigate how teachers developed learning materials of interdisciplinary nature and system approach and elements of new modes of assessment, how they viewed the implementation of new modes of assessment in their classrooms, and how these methods could be applied to assess the teachers' deliverables (Dori & Herscovitz, 2000).

6.2 Research Setting

The research population included about 50 teachers, 60% of whom came from the Jewish sector and 40% from the Arab sector. About 80% of the population were women, 65% were biology teachers, and the rest were chemistry, physics, or technology teachers. About 67% of these science and technology teachers had over 10 years teaching experience.

During the three years of their professional development, the junior-high science teachers were exposed to several science topics in the workshops. Scientific, environmental, societal, and technological aspects of these topics were presented through laboratory experiments, case studies and cooperative learning. During the first two years, teachers were required to carry out three projects. The assignments included choosing a topic related to science and technology, which was not covered in the workshops. While applying system approach, the teachers had to develop a case study and related student activities as part of the project.

The first project, "Elements," which was carried out toward the end of the first year, concerned a case study on a chemical element taken from a popular science journal. The teachers got the article and were asked to adapt it to the students' level and design student activity, which would follow reading it.

The second project, "Air Pollutants", was carried out during the middle of the second year. Here, teachers were required to search for an appropriate article that discussed this topic and dealt with a scientific/technological issue. Based on the article they selected, they had to design a case study along with an accompanying student assignment.

The third and final project, which started toward the end of the second year, included preparing a comprehensive interdisciplinary teacher-chosen subject, designing a case study and student activities, and implementing it in their classes. The first, second and third projects were done individually, in pairs and in groups of three to four teachers, respectively. The third project was taught in the teachers' own classrooms, and was accompanied by peer and teacher assessment, as well as students' feedback.

6.3 Assessment

The individual teacher, peers and the workshop lecturer assessed the first projects. The objective of this assessment was to experience the use of new modes of assessment. In the second project, each pair presented their work orally to the entire group, and the pair, the other pairs and the lecturer assessed it. The third project was presented by each group in an exhibition

A Framework for Project-Based Assessment in Science Education

and was evaluated using the same criteria and the same assessment scheme and method as the two previous projects.

Setting criteria for assessing the projects preceded the new modes of assessment that the science teachers applied. Groups of 3-4 teachers set these criteria after reading their peer project's portfolios. Six criteria were finally selected in a plenary session. Some of these criteria, such as design/aesthetics, and originality/creativity, were concerned with project assessment in general, and were therefore also applicable to students' projects. Other, more specific criteria related to the assessment of the teacher portfolios, and included interdisciplinarity, suitability for the students, and variability of the accompanying activities.

The originality/creativity criterion was controversial. While most groups proposed a criterion that included these elements, it was apparent that objective scoring of creativity is by no means a straightforward matter. One group therefore suggested that this criterion would add a bonus to the total score. Teachers were also concerned about the weight assigned to each criterion. The decision was that for peer assessment during the workshops, all criteria would be weighted equally, while for classroom implementation, the teacher would have the freedom to set the relative weights after discussing it with his/her students.

6.4 Method of Analysis and Findings

The criteria proposed by the teachers, along with additional new ones, were used to analyse both the case study and the accompanying activities that teachers had developed.

The analysis of the case study was based on its level of interdisciplinary nature and system approach, as well as the suitability to students' thinking skills. Two science and environmental education experts validated the classification and analysis of the case studies and the related activities. Analysing the case studies teachers developed, we found that they went through a change from viewing only their own discipline to a system approach that integrates different science disciplines. The level of suitability of the case study to the target population increased accordingly. The statistical mode of the level of interdisciplinarity (number of disciplines integrated) in the case studies increased from one in the first project (with frequency of 50%) to two in the second project (50%) and to three in the third (80%). In parallel, the suitability for students increased from low (42%) through intermediate (37%) to high (60%).

For the **student activity's assessment** we used four categories: (1) interdisciplinarity; (2) variety; (3) relation to the case study; and (4) complexity (Herscovitz & Dori, 2000). The score for each criterion in each category for assessing the student activities that followed the case study are presented in Table 6.

Table 6. Categories and scores for assessing case study student activities

Category	Points	Score for each criterion
Level of inter-disciplinary nature	1 2 3	One domain only (scientific, societal, etc.) is involved. Two domains are involved. Three domains are involved.
Variability	1 +1	All activities are of the same type: teacher questions to the student. An extra point is given for each additional activity (e.g., experiment, movie, concept map, class discussion, field trip, debate).
Relation to the Case Study	1 2 3	Low: superficial treatment, which does not touch the essence of the problem; activity has little to do with the case study. Intermediate: reasonable treatment and relation to the case study; no deep treatment of the problem raised in the case study. High: deep, serious treatment of the case study; gradual, logical construction that leads to profound student understanding.
Complexity	1 2 3	Low-order thinking skill: the answer is contained in the case study. It requires knowledge and understanding only. High-order thinking skill: the answer is, at most, partially contained in the case study. It requires analysis and synthesis. Very high-order thinking skill: the answer is not contained in the case study. It requires value judgement, system approach, argumentation, or assessment.

A Framework for Project-Based Assessment in Science Education 111

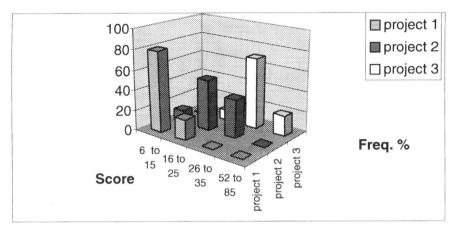

Figure 2. Total assessment of case study activities in the three projects

Figure 2 shows a clear trend of improvement of the total score of case study activities in each one of the three projects. The most frequent score range in project 1 was 6 to 15 (with frequency of 80%), in project 2 – 16 to 25 (50%), and in project 3 – 26 to 35 (70%). In project 3, 20% of the works were ranked in the range of 52 to 85. No work in project 1 or 2 was ranked in this range. The grades for the variability of the accompanying activities increase as well.

Using the criteria the teachers had set, they assessed their own project, as well as their peers', providing only verbal remarks without assigning numerical values. Teachers' opinions towards performance of peer assessment in class met with decreasing resistance as the projects progressed. For most teachers, the criteria-setting process was a new, inspiring experience. One teacher noted: *"I had heard about student and peer assessment, but I had no idea what it entails and how it should be implemented. Now I know that I need to involve my students in setting the criteria."* Another teacher indicated: *"I had hard time explaining to my students why one student portfolio got a higher score than another. Thanks to the discussion during the workshop, the issue of criteria setting became clear... Involving students in setting the criteria and in assessing their own work as well as their peers', fosters involvement and enhances the collaborative aspect of their work."* A third teacher said that the discussion with her peers about the new modes of assessment contributed to her pedagogical knowledge: *"I realized that I need to add knowledge questions in student activities following the case study, so that the low academic level students can participate in group discussions and demonstrate their point of view."*

The teachers were enthusiastic about these new modes of assessment, in which teachers and students may play the role of equal partners. Some expressed readiness to implement this approach in their classrooms and indeed invited the lecturer to observe them in action.

Teachers developed learning materials with an increased level of system approach and suitability to their students. They incorporated some new elements of assessment into the activities that followed the case studies. The assessment of teachers' projects has shown that the activities they proposed towards the end of the second year increased in complexity and required the students to exhibit higher order thinking skills, such as argumentation, value judgment, and critical thinking. It should be noted that these results involve several factors, including individual vs. group processes, long-term (two-year) learning, assessment methods, and relative criteria weight in peer- and classroom assessments. Future studies may be able to address each of these factors separately, but as experience has shown this is hardly feasible in education.

7. DISCUSSION

The project-based assessment framework has emerged as a common thread throughout the three studies described here. This framework is holistic, in the sense that it touched upon domains, activities and aspects that both students and teachers experience. Researchers and educators attribute many benefits to project-based assessment schemes. Among these benefits are the generation of valid and reliable information about student performance, provision of formative functions, and promotion of teacher professionalism (Black, 1998; Nevo, 1994; Dori & Tal, 2000; Worthen, 1993). As Sarasor (1990) wrote, the degree of responsibility given to students in the traditional classroom is minimal. They are responsible only in the sense that they are expected to complete tasks assigned by teachers and do so in ways the teachers have indicated. They are not responsible to other students. They are solo learners and performers responsive to one adult. The opposite is true for the project-based framework, into which the new modes of assessment were woven in a way that constitutes a natural extension of the learning itself rather than an external add-on. In establishing the assessment model, we adopted lessons of Nevo (1995) in school-based evaluation, who noted that outcomes or impacts should not be the only thing examined when evaluating a program, a project, or any other evaluation object within school.

A Framework for Project-Based Assessment in Science Education 113

Departing from the traditional school life-learning environment, the project-based approach resembles real life experience (Dori & Tal, 2000; Mitchell, 1992a). Students are usually enthusiastic about learning in project-based settings, they apply inquiry skills and deal with complexity while using methods of scaffolding (Krajcik, Blumenfeld, Marx, Bass, Fredricks, & Soloway, 1998). In the project-based learning described in this chapter, students were responsible for their own learning, teachers oversaw student teamwork, and community stakeholders were involved in school curriculum and assessment. Participants eagerly engaged in the learning process with emotional involvement, resulting in meaningful and long-term learning. In a school environment like this, higher order thinking skills and autonomous learning skills develop to a greater extent than in traditional learning settings (Dori, 2003).

Black (1995, 1998) argued that formative assessment has much more potential than usually experienced in schools, and that it affects learning processes in a positive way. The assessment types presented in this chapter as the project-based framework increases the variety of assessment models. One advantage of this type of educational orientation is that students and teachers collaborate to create a supportive learning environment, which is in-line with the knowledge building in a community of learners (Bereiter & Scardamalia, 1993). The project-based curriculum and its associated assessment system required time and effort investment by the teachers and students alike. Yet they accepted it, as they recognized the value of the assessment as an on-going process, integrated with the learning.

The project-based assessment framework that has emerged from the studies presented here is multi-dimensional in a number of ways:
- The assessed objects are both the individual student (or teacher) and the team;
- External experts, teachers, and students carry out the assessment;
- The assessing tools are case studies, projects, exhibition, portfolios and self-assessment questionnaires.

Despite its complexity, this assessment was meaningful and suitable to a variety of population types.

In all the three project-based studies, students achieved scores in the high-level assignments that were lower than those achieved in the low-level assignments. This is consistent with finding of other researchers (Harlen, 1990; Lawrenz et al., 2001) who showed that open-ended assignments are more difficult and demanding, because they measure more higher order thinking skills and because students are required to formulate original responses. Open-ended, high-level assignments provide important feedback that is fundamentally different in nature than what can be obtained from assignments that are defined as low-level ones. The high level assignments, developed as research instruments for these studies, required a variety of higher order thinking skills and can therefore serve as a unique diagnostic tool.

Following the recommendation of Gitmore and Duschl (1998) and of Treagust et al. (2001), in the "Matriculation 2000" Project teachers improved students' learning outcomes and shaped curriculum and instruction decisions at the school and classroom level through changing the assessment culture. The reform that took place in the 22 high schools is a prelude to a transition from a nationwide standardized testing system to a school-based assessment system. Moreover, teachers, principals, superintendents, and Ministry of Education officials who were engaged in this Project became involved in convincing others to extend the Project boundaries to additional courses at the same school and to additional schools in the same district.

The study that involved teachers has shown that projects can serve as a learning and assessment tool not only for students but also for teachers. Hence, incorporating project-based assessment is recommended for both pre- and in-service teacher workshops. Introducing teachers to this method will not only serve as assessment means to evaluate the teachers, but will also expose them to new modes of assessment and encourage them to implement it in their classes. Relevant stakeholders in the Israeli Ministry of Education have recognized the significance of the research findings of these studies and others carried out in other academic institutes. They realize the value of project-based learning and the new modes of assessment framework. However, economical constrains have been slowing down its adoption on a wider scale.

The main limitation of these studies stems from the scale up problem. It is difficult to implement project-based assessment with great numbers of learners. If we believe that assessment modes as described in this chapter ought to be applied in educational frameworks, we need to find efficient ways to alleviate teachers' burden of following, documenting and grading students' project portfolios. The educational system should take care of adequate compensation arrangements that would motivate teachers to carry on these demanding assessment types even after the initial enthusiasm has diminished. Pre-service teachers can be of significant help in this regard.

The findings of these studies clearly indicate that project-based assessment when embedded throughout the teaching process has the unique advantage of fostering and assessing higher order thinking skills. These conclusions warrant validation through additional studies in different settings and in various countries.

REFERENCES

Bakshi, T. S., & Lazarowitz, R. (1982). A Model for Interdisciplinary Ecology Project in Secondary Schools. *Environmental Education and Information, 2* (3), 203-213.

Baxter, G. P., Shavelson, R. J., Goldman, S. R., & Pine, J. (1992). Evaluation of procedure-based scoring for hands-on science assessment. *Journal of Educational Measurement, 29*, 1-17.

Baxter, G. P., & Shavelson, R. J. (1994). Science performance assessments: Benchmarks and surrogates. *International Journal of Educational Research, 21*, 279-297.

Ben-Elyahu, S. (1995). *Summary of the feedback questionnaire of "Matriculation 2000" Project first year*. Pedagogical Secretariat, Ministry of Education, Jerusalem, Israel (in Hebrew).

Ben-Peretz, M. (1994). *Report of the Committee for Examining the Format of Israeli Matriculation Examination*, Ministry of Education, Jerusalem, Israel (in Hebrew).

Bereiter, C., & Scardamalia, M. (1993). Schools as Nonexpert Societies. In: *Surpassing Ourselves An Inquiry Into The Nature and Implications of Expertise*, pp. 183-220. Chicago: Open Court.

Birenbaum, M., & Dochy, F. J. R. C. (Eds.). (1996). *Alternatives in assessment of achievements, learning processes and prior knowledge*. Boston, MA: Kluwer.

Black, P. (1995). Curriculum and assessment in science education: the policy interface. *International Journal of Science Education, 7*, 453-469.

Black, P. (1995a). Assessment and Feedback in Science Education. *Studies in Educational Evaluation, 21*, 257-279.

Black, P. (1998). Assessment by Teachers and the Improvement of Student's Learning. In: Fraser, B. & Tobin, K (Eds.), *International Handbook of Science Education* (pp. 811-822). Kluwer Academic Pub.

Blumenfeld, P. C., Marx, R. W., Soloway, E., & Krajcik, J. (1996). Learning with Peers: From Small Group Cooperation to Collaborative Communities. *Educational Researcher, 25* (8), 37-40.

Cheung, D., Hattie, J., Bucat, R., & Douglas, G. (1996). Measuring the Degree of Implementation of School-based Assessment Schemes for Practical Science. *Research in Science Education, 26* (4), 375-389.

Costa, A. L. (1985). Teacher behaviors that enable student thinking. In A. L. Costa (Ed.), *Developing minds: a resource book for teaching thinking*. Alexandria, Va: Association for Supervision and Curriculum Development.

Darling-Hammond, L. (1994). *Standards for Teachers*. Paper Presented at the Annual Meeting of the American Association of Colleges for Teacher Education, Chicago, IL, ED378176.

Dillon, J. T. (1990). *The practice of questioning*. London: Routledge.

Dori, Y. J. (1994). Achievement and Attitude Evaluation of a Case-Based Chemistry Curriculum for Nursing Students. *Studies in Educational Evaluation, 20* (3), 337-348.

Dori, Y. J. (2003) From Nationwide Standardized Testing to School-based Alternative Embedded Assessment in Israel: Students' Performance in the "Matriculation 2000" Project. *Journal of Research in Science Teaching, 40* (1).

Dori, Y. J., Barnea, N., & Kaberman, Z. (1999). *Assessment of 22 High School in the "BAGRUT 2000" (Matriculation 2000) Project*. Research Report for the Chief Scientist, Israeli Ministry of Education. Department of Education in Technology and Science, Technion, Haifa, Israel (in Hebrew).

Dori, Y. J., & Herscovitz, O. (1999). Question Posing Capability as an Alternative Evaluation Method: Analysis of an Environmental Case Study. *Journal of Research in Science Teaching, 36* (4), 411-430.

Dori, Y. J., & Herscovitz, O. (2000). *Project-based Alternative Assessment of Science Teachers.* Paper presented at the 1st Biannual Conference of the EARLI Assessment SIG – "Assessment 2000", University of Maastricht, Maastricht, The Netherlands.

Dori, Y. J., & Tal, R. T. (2000). Formal and informal collaborative projects: Engaging in industry with environmental awareness. *Science Education, 84,* 1-19.

Flores, G., & Comfort, K. (1997). Gender and racial/ethnic differences on performances assessments in science. *Educational Evaluation and Policy Analysis, 19* (2), 83-97.

Gallard, A. J., Viggiano, E., Graham, S., Stewart, G., & Vigiliano, M. (1998). The learning of vouluntary and involuntary minorities in science classrooms. In B. J. Fraser & K. G. Tobin (Eds.), *International Handbook of Science Education* (pp. 941-953). Kluwer Academic Publisher, Dordrecht/Boston/London.

Gitmore, D. H., & Duschl, R. A. (1998). Emerging issues and practices in science assessment. In B. J. Fraser & K. G. Tobin (Eds.), *International Handbook of Science Education* (pp. 791-810). Kluwer Academic Publisher, Dordrecht/Boston/London.

Guba, E. G., & Lincoln, Y. S. (1989). *Fourth generation evaluation.* London: Edward Arnola Publishers, Ltd.

Harari, H. (1994). *Tomorrow 98: report of the superior committee on science, mathematics and technology education of Israel.* Jerusalem: State of Israel, ministry of education, culture and sport.

Harlen, W. (1990). Performance testing and science education in England and Wales. In A. B. Champagne, B. E. Lovitts & B. J. Calenger (Eds.), *Assessment in the Service of Instruction* (pp. 181-206). Washington, DC: American Association for the Advancement of Science.

Herried, C. F. (1994). *Case studies in science - A novel method of science education,* (4), 221-229.

Herried, C. F. (1997). What is a case? Bringing to science education the established tool of law and medicine. *Journal of College Science Teaching, 27,* 92-95.

Herscovitz, O., & Dori, Y. J. (2000) *Science Teachers in an Era of Reform – Toward an Interdisciplinary Case-based Teaching/Learning.* Paper presented at the NARST Annual Meeting – the National Association for Research in Science Teaching Conference, New Orleans, LA, USA.

Hofstein, A., & Rosenfeld, S. (1996). Bridging the Gap Between Formal and Informal Science Learning. *Studies in Science Education, 28,* 87-112.

Keiny, S. (1995). *STES Curriculum Development as a Process of Conceptual Change.* A Paper Presented at NARST Annual Meeting, San Francisco CA.

Krajcik, J., Blumenfeld, P. C., Marx, R. W., Bass, K. M., Fredricks, J. F., & Soloway, E. (1997). Inquiry in Project-Based Science Classroom: Initial Attempts by Middle School Students. *The Journal of the Learning Sciences, 7* (3/4), 313-350.

Lawrenz, F., Huffman, D., & Welch, W. (2001). The science achievement of various subgroups on alternative assessment formats. *Science Education, 85,* 279-290.

Lazarowitz, R., & Hertz-Lazarowitz, R. (1998). Cooperative Learning in the Science Curriculum. In B. Fraser & K. Tobin (Eds.), *International Handbook of Science Education* (pp. 449-469). Dordrecht, Netherlands: Kluwer Academic Publishers.

Leeming, F. C., Dwyer, W. O., & Bracken, B. A. (1995). Children's Ecological Attitude and Knowledge Scale (CHEAKS): Construction and Validation. *Journal of Environmental Education, 26,* 22-31.

Lewy, A. (1996). Postmodernism in the Field of Achievement Testing. *Studies in Educational Evaluation, 22* (3), 223-244.

McDonald, J., & Czerniac, C. (1994). Developing Interdisciplinary Units: Strategies and Examples. *School Science and Mathematics, 94* (1), 5-10.

Mitchell, R. (1992a). *Testing for Learning: How New Approaches to Learning Can Improve American Schools*. New York: The Free Press.

Mitchell, R. (1992b). Getting Students, Parents and the Community into Act. In: *Testing for Learning: How New Approaches to Learning Can Improve American Schools* (pp. 79-101). New York: The Free Press.

Nevo, D. (1983). The Conceptualization of Educational Evaluation: An Analytical Evaluation of the Literature. *Review of Educational Research, 53*, 117-128.

Nevo, D. (1994). Combining Internal and External Evaluation: A Case for School-Based Evaluation. *Studies in Educational Evaluation, 20* (1), 87-98.

Nevo, D. (1995). *School-Based Evaluation: A Dialogue for School Improvement*. Oxford, GB: Elsevier Science Ltd, Pergamon.

NRC - National Research Council (1996). *National Science Education Standards*. Washington DC: National Academy Press.

Pond, K., & Ul-Haq, R. (1997). Learning to Assess Students using Peer Review. *Studies in Educational Evaluation, 23* (4), 331-348.

Posch, P. (1993). Research Issues in Environmental Education. *Studies In Science Education, 21*, 21-48.

Resnick, L. B. (1987). *Education and learning to think*. Washington D.C.: National Academy Press.

Resnick, L. B., & Resnick, D. P. (1992). Assessing the thinking curriculum: New tools for educational reform. In B. R. Gifford & M. C. O'Connor (Eds.), *Changing assessments: Alternative views of aptitude, achievement and instruction* (pp. 37-75). Boston: Kluwer Academic Publishers.

Ruiz-Primo, M. A., & Shavelson, R. J. (1996). Rhetoric and reality in science performance assessment: An update. *Journal of Research in Science Teaching, 33*, 1045-1063.

Sarason, S. (1990). *The predictable Failure of Educational Reform*. San Francisco: Jossey-Bass.

Shavelson, R. J., & Baxter, G. P. (1992). What we've learned about assessing hands-on science. *Educational Leadership, 49*, 20-25.

Shepardson, D. P., & Pizzini, E. L. (1991). Questioning levels of junior high school science textbooks and their implications for learning textual information. *Science Education, 75*, 673-682.

Shorrocks-Taylor, D., & Jenkins, E. W. (2000). *Learning from others*. Kluwer Academic Publishers, Dordrecht, The Netherlands.

Shulman, L. S. (1986). Those Who Understand: Knowledge Growth in Teaching. *Educational Researcher, 15* (2),4-14.

Sizer, T. (1992). *Hoace's School: Redesigning the American High School*. Boston: Houghton Mifflin.

Solomon, J. (1993). *Teaching Science, Technology and Society*. Philadelphia: Open University Press.

Sweeney, A. E., & Tobin, K. (Eds.). (2000). *Language, discourse, and learning in science: Improving professional practice through action research*. Tallahassee, Fl:SERVE.

Tal, R., Dori, Y. J., & Lazarowitz, R. (2000). A Project-Based Alternative Assessment System. *Studies in Educational Evaluation, 26, 2*, 171-191.

Tamir, P. (1998). Assessment and evaluation in science education: Opportunities to learn and outcomes. In B. J. Fraser & K. G. Tobin (Eds.), *International Handbook of Science Education* (pp. 761-789). Dordrecht/Boston/London: Kluwer Academic Publisher.

Treagust, D. F., Jacobowitz, R., Gallagher, J. J., & Parker, J. (2001). Using assessment as a guide in teaching for understanding: A case study of a middle school science class learning about sound. *Science Education, 85*, 137-157.

Wiggins, G. (1989). A true test: Toward more authentic and equitable assessment. *Phi Delta Kappan, 70*, 703-713.

Worthen, B. R. (1993). Critical Issues That Will Determine the Future of Alternative Assessment. *Phi Delta Kappan, 74* (6), 444-457.

Zohar, A., & Dori Y. J. (2003). Higher Order Thinking Skills and Low Achieving Students – Are they Mutually Exclusive? *Journal of the Learning Sciences, 12* (3).

Zoller, U. (1991). Problem Solving and the "Problem Solving Paradox" in Decision-Making-Oriented Environmental Education. In S. Keiny & U. Zoller (Eds.), *Conceptual Issues in Environmental Education* (pp.71-88). New York: Peter Lang.

Evaluating the OverAll Test: Looking for Multiple Validity Measures

Mien Segers
Department of Educational Development and Research, University Maastricht, The Netherlands

1. INTRODUCTION

It is widely accepted that to an increasing extent, successful functioning in society demands more than being capable of performing the specific tasks a student learned to perform. Society is characterized by continuous, dynamic change. This has led to major shifts in the conception of the aim of education. Bowden & Marton (1998) expressed this change as the move from "learning what is known" towards "educating for the unknown future". As society changes rapidly, there will be a growing gap between what we know at this moment and what will be know in the coming decade. Within this context, what is the sense for students to consume encyclopaedic knowledge? Bowden & Marton advocate that cognitive, meta-cognitive and social competencies are required, more than before. Birenbaum (1996, p. 4) refers to cognitive competencies such as problem solving, critical thinking, formulating questions, searching for relevant information, making informal judgments, efficient use of information, etc. The described changes are taking place with increasing moves towards what is called powerful learning environments (De Corte, 1990). They are characterized by the view that learning means actively constructing knowledge and skills based on prior knowledge, embedded in contexts that are authentic and offer many opportunities for social interaction. Feltovich, Spiro, and Coulson (1993) use the concept of understanding to describe the main focus of the current instructional and assessment approach. They define understanding as acquiring and retaining a network of concepts and principles about some domain that accurately represents key phenomena and their interrelationships and that can be engaged flexibly when pertinent to accomplish diverse, sometimes novel objectives." (p. 181). Examples of powerful learning environments include problem-based learning, project-oriented learning and product-oriented learning.

With respect to assessment, a variety of new modes of assessment are implemented, with performance assessment and portfolio-assessment as two well-known examples. Characteristic of these new modes of assessment is their "in context" and "authentic" nature. Authentic refers to the type of cognitive challenges in the assessment. Assessment tasks are defined as authentic when "the cognitive demands or the thinking required are consistent with the cognitive demands in the environment for which we are preparing the learner" (Savery & Duffy, 1995, p.33). However, authentic not only refers to the nature of the assessment tasks, but also to the role of assessment in the learning process. The assessment tools are tools for learning. Teaching and assessment are both seen as tools to support and enhance students' transformative learning. Assessment is a valuable learning experience in addition to allowing grades to be assigned (Birenbaum & Dochy, 1996; Brown & Knight, 1994). In this respect, the formative function of assessment is stressed. Finally, the shift from teacher-centred to student-centered education has its impact in the division of responsibilities in the assessment process. To a growing extent, the student is an active participant in the evaluation process, who shares responsibility, practices self-evaluation, reflection and collaboration.

As the criteria for assessment have changed, questions have been raised about the conceptions of quality criteria. As described in chapter 3, conceptions of validity and reliability encompassed within the instruments of assessment, have changed accordingly. One of these changes is the growing accent on the consequential validity of assessment. However, until now, only a few studies explicitly address this issue in the context of new modes of assessment. This chapter will present the results of validity studies of the OverAll Test, indicating the value-added of investigating the consequential validity of new modes of assessment.

2. THE OVERALL TEST AND THE PROBLEM BASED LEARNING ENVIRONMENT

The OverAll Test is a case-based assessment instrument, assessing problem-solving skills. With the implementation of the OverAll Test, it was expected that the alignment between curriculum goals, instruction and assessment would be enhanced. It was expected that by assessing students skills in identifying, analysing, solving and evaluating novel authentic problems, it would stimulate the problem-solving process in the tutorial groups. Now, it has been implemented in a wide variety of faculties at different Belgian and Dutch institutions of higher education. Most of them have adopted a problem-based or project based instructional approach and they have attempted to optimise the learning effects of their instructional innovations by changing the assessment practices.

2.1 The Problem-Based Learning Environment

One of the aims of problem-based learning is to educate students who are able to analyse and solve problems (Barrows, 1986; Engel, 1991; Poikela & Poikela, 1997; Savery & Duffy, 1995). Therefore, the learning process is initiated and guided by a sequence of a variety of problem tasks, which cover the subject content (Nuy, 1991). During the subsequent years of study, these problem situations become more complex and diverse in activities to be undertaken by the students (e.g. from writing an advice for a simulated manager in problems to discussing in a live setting the proposal with a manager from a specific firm). Working in a small group setting (10 to12 students), guided by a tutor, students analyse the problem presented and discuss the relevant aspects of the problem. They formulate a set of learning objectives based on their hypotheses about possible conceptualisations and solutions. These objectives are the starting point for students in order to process the subject matter in study books. In the next group sessions, the findings of the self-study activities are reported and discussed. Their relevance for the problem and for novel but similar problems is evaluated.

The curriculum consists of a number of instructional periods, called blocks. Each of them has one central theme, operationalized in a set of problems.

2.2 The OverAll Test

The OverAll Test is a case-based assessment instrument. The cases are not of the key-feature format (Des Marchais & Dumais, 1993) but describe the problem situation in an authentic way, i.e. with for the problem situation relevant and irrelevant elements. This implies that for most cases, in order to understand and analyse the problem situation, knowledge from different disciplines has to be mastered. The test items require students to define the problem, analyse it, contribute to its solution and evaluate the solutions. They do not ask to tackle the entire problem situation presented in each case but refer to its critical aspects. The cases present novel problems, asking the students to transfer the knowledge and skills they acquired during the tutorials and to demonstrate the understanding of the influence of contextual factors on problem analysis as well as on problem solving. For some test items, students are asked to argue their ideas based on various relevant perspectives

For generalizability purposes (Shavelson, Gao, & Baxter 1996), the OverAll Test items refer to a set of cases.

In summary, the OverAll Test can be described by a set of characteristics (Segers, 1996):
- it is a paper-and-pencil test;
- it is part of the final examination;
- it presents a set of authentic cases that are novel to the students (this means they were not discussed in the tutorial groups);
- the test items require from the students that they identify, analyse, solve and evaluate the problems underlying the cases;
- the cases are multidisciplinary;
- two item formats are used: multiple-choice questions and open-ended questions.
- it has an open-book character;
- the test items refer to approximately seven different authentic cases (each about 10 to 30 pages) that are available for the students from the beginning of the instructional period; the test items related to the cases are given at the moment of test administration.

Figure 1 presents examples of OverAll Test items. The test items refer to the Mexx and Benetton case (30 pages). The case study presents the history and recent developments in the fashion companies Mexx and Benetton. Main trends within the European clothing industry are described. Mexx Fashion and Benetton as a company are pictured by the organisational structure, the product profile and market place, the business system, the corporate culture and some actual facts and figures.

The first question refers to the different viewpoints on management. Memorisation of the definition is not sufficient to answer the OverAll Test item. Students have to interpret the case and select the relevant information for this test item. On the basis of a comparison of this information with the conceptual knowledge of the different viewpoints on management, they have to deduce the answer. The second OverAll Test item refers to the concept of vertical integration. It requires students to take a set of mental steps to reach the solution of the problem posed. For the first part of the question (a), these can be schematised as follows:

1. Define the concept of corporate strategies.
2. Select the relevant information for the Mexx company as described in the case study.
3. Confront it with the definition of the different possible strategies.
4. Select the relevant information for Benetton as described in the case study.
5. Apply/choose a definition of the different possible strategies.
6. Match the relevant information of both cases with the chosen definition of the strategies.
7. Define for each company its strategy.
8. Compare both strategies by going back to the definition of the strategies and the relevant information in the case study.

Case Mexx/Benetton
(This case is approx. 35 pages)

The case study presents the history and recent developments in the fashion company Mexx. Main trends within the European clothing industry are described. Mexx Fashion and Benetton are pictured by their organizational structure, the product profile and marketplace, the business system, the corporate culture and some actual facts and figures.

Question 1
true/?/false Mexx's corporate culture and philosophy is consistent with the systems viewpoint on management.
(False, it is consistent with behavioural viewpoint)

Question 2
Benetton's and Mexx's corporate strategies are quite different. More specifically, there are two main differences.
a. Identify these two main differences in corporate strategies. Illustrate your answer with examples mentioned in the case.
b. What are the advantages of Benetton's corporate strategy compared to Mexx's approach?

Figure 1. Examples of OverAll Test Questions

For the second part (b), students have to evaluate. Therefore, they have to take some extra mental steps:
1. Understand the conditions to be efficient and effective for the different strategies.
2. Select the relevant information on the conditions for both companies.

3. Interpret the actual conditions by comparison with those studied in the textbooks.

This example illustrates the OverAll Test measures if students are able to retrieve the relevant concept (model, principles) for the problem. Furthermore, it measures if they can use these instruments for solving the problem. They measure if the knowledge is usable (Glaser, 1990) or if they know "when and where" (conditional knowledge). In short, the OverAll Test measures to what extent students are able to analyse problems and contribute to their solution by applying the relevant instruments. Figure 2 presents a second example of OverAll Test items based on an article on Scenario Planning.

Question 1
In his introduction, Schoemaker (1995) compares the method of scenario planning with other approaches such as contingency planning, sensitivity analysis and computer simulations. Hellriegel & Slogum (1996) textbook) give a similar comparison of three methods: scenarios, the Delphi technique and simulation. They stress that an overlap exists between these approaches, and indeed, it is not difficult to imagine how to use techniques like Delphi and simulation within Schoemaker's framework of scenario planning.

True/?/false   The Delphi technique fits better in phase 3 of scenario planning (identifying basic trends) than in phase 9 of the scenario planning (develop quantitative models).

Question 2
The correlations in Table 3, Part B on p. 31 (Schoemaker, 1995) are nearly all positive, which makes the case rather specific.......Give a new example of scenario planning by solving the following tasks:
1. Write down a hypothetical correlation matrix, having the same size as that in table 3, but the number of entries with +, - and 0 more equally distributed
2. Derive a scenario profile (as figure 1) from this correlation matrix, thereby given special attention to the existence of both positive and negative correlations. If necessary, make additional assumptions in order to find the profile. Start with one single scenario.
3. Derive a second scenario profile, assuming this second scenario to be the reverse scenario of the first (reverse in its literal meaning; if the first scenario is something like recession, then the second is that of high economic activity).
4. Give a description in words of the consistency requirements one has to observe in assignments b and c.
5. Give an interpretation of the outcomes of the scenario profiles you constructed yourself. Schoemaker ends up with one scenario that performs best in all possible aspects, whilst the third scenario is the worst one, again in all possible aspects. Is the same true for the case you designed?

Figure 2. Examples of OverAll Test questions

The Schoemaker test items ask to address the problem from different perspectives. They have to integrate knowledge of the disciplines of Statistics and Organisation and this within the context of scenario planning. Knowledge from both disciplines has to be used to tackle the problem of scenario planning.

3. THE VALIDITY OF THE OVERALL TEST

In 1992, a research project started aiming to monitor different quality aspects of the OverAll Test. One of the studies addressed the instructional validity of the OverAll test. With the changing notions of validity and based on different problems observed within the faculty, in 1999, a study was conducted to measure the consequential validity of the OverAll test. Both studies will be presented in the next session.

3.1 The Instructional Validity of the OverAll Test

McClung (1979) introduced the term "instructional validity" as opposed to the term "curricular validity". As an answer to the question "Is it fair to the students to answer the assessment tasks?", assessment developers mostly check the curricular validity, i.e. the match between the assessment content and the formal curriculum. The description of the formal curriculum is derived from the curricular objectives and the curriculum content. The formal curriculum is mostly expressed in the blueprint of the curriculum, describing the educational objectives in terms of content and level of comprehension. The assessment objectives are expressed in the assessment blueprint that is mirrored in the curricular blueprint. In interpreting test validity based on the match between the curriculum and the assessment blueprint, it is assumed that the instructional practice reflects the formal curriculum. Many studies indicate that this assumption may be questioned (Calfee, 1983; De Haan, 1992; English, 1992; Leinhardt & Seewald, 1981; Pelgrim, 1990). The operational curriculum, defined as what is actually learned and taught in the classrooms, can significantly differ from the formal curriculum as described in textbooks and syllabi. This is especially the case in learning environments were, more than in more teacher-centered settings, students are expected to formulate their own learning goals. How sure can we be that the test is fair to students who vary to some extend in the learning goals they pursue?

Only a few studies in student-centred learning environments, such as problem-based learning, addressed this issue. Dolmans (1994) investigated the match between formal and operational curriculum and the relation between attention given to learning goals during the tutorial groups and the students' assessment outcomes. She concluded there was an overlap of 64.2% (s= 26.7) between both curricula. The correlation between the time spent on the core concepts and the test items referring to these concepts, seemed to be significant but weak ($r=.22$, $p<0.5$, $n=94$). Probably it is the quality, more than the quantity, of the time spent on the core concepts that affects test scores.

3.1.1 Research Method

3.1.1.1 Procedure

The formal curriculum was described by analysing the textbooks, syllabi and tutorial manuals. The analysis resulted in a list containing more than 500 detailed topics for each period. This extended list was screened by domain specialists to get a workable list. They constructed a hierarchical schema of the list of topics. The highest hierarchical levels of the networks of subjects are included in the final version; for example the concepts of "entry strategies", "export", "licensing" and "joint ventures". They were all included in the draft version. In the final version only the concept of "entry strategies" was included. Thus, the list of central concepts was reduced to 147 topics for the Marketing and Organization period and 136 topics for the Macro-economics period. The curricular validity is examined by comparing the formal curriculum with the test of the first instructional period. The list of concepts is compared with the list of objectives of the OverAll Test.

To examine the instructional validity, two questionnaires were developed based on the lists of concepts. The questionnaires are a modified version of the Dolmans Topic Checklist (1994). The first Topic Checklist (TOC1) consists of the 147 topics in the disciplines Marketing and Organization. The TOC2 presented the 136 Macroeconomics topics.

Students were asked to indicate whether the topic was discussed in their tutorial groups or not, by marking the topic or not. In order to gain some insight into the quality of the time spent on the topic, the second Topic Checklist (TOC 2) on Macroeconomics consisted of two additional questions. Students had to indicate the level of comprehension they believed they had reached. For every respondent, the number of topics mastered on each of the three levels of comprehension was counted. These levels were defined as the level of definition, the level of comprehension and the level of analysis. Mastery on the level of definition indicates the student is (only) able to reproduce the meaning of the concept as formulated in the textbooks. Comprehension of the topic implies that the student is able to define the concept in his own words, describe its relevance and its relation to other concepts. To master a topic on the level of analysis would require the student to be able to apply the concepts when being confronted with a problem to be analysed. The staff members who developed the course were asked to indicate for each topic the intended level of comprehension. Finally, students were asked if a topic received much, moderate, or not much attention during the tutorial meetings.

3.1.1.2 Sample

The sampling procedure employed in the study was that of the quota sample. The group of first year students was, for organizational reasons, divided into four groups. Two groups had their meetings in the morning, two groups in the afternoon. Students were equally selected from these four groups. For the TOC 1, 34 students participated voluntarily, for TOC 2, 45 students.

3.1.2 Results

As the results in Table 1 indicate, there is an important amount of overlap between topics planned for study by the staff, and the topics indicated by the students as being subject of discussions and study during the instructional period. Table 1 indicates that on average 87% of the topics of TOC 1 and 77.4% of the topics of TOC 2 have been subject of study (RT).Other studies investigating the match between the formal and the operational curriculum in a Problem-Based Learning-setting (Dolmans, 1994), show an overlap of 64.2%.

Students perceived they had mastered on average 47% of the topics of TOC 2 on the level of comprehension, i.e., that they were able to explain in their own words the meaning of the topics, their relevance and their relation to other concepts. For an average of 31% of the topics, students stated they were able to use these topics for the analysis of problems (level of analysis). For 22% of the topics, on average, students indicated they had mastered them on the level of definition, i.e., that they were able "only" to reproduce the definition. The correspondence with the aims of the staff is considerable.

Concerning the curricular validity, for the OverAll Test, 11% and 15% of the test items refer to topics that were not part of the formal curriculum. This means they were missing in the Topic Checklist I. Comparing topics that had either been or not been discussed (RT/NRT) with test items content, none of the topics which were indicated as not having been subject of the discussions by more than 29% of the students (percentile 25) were part of the OverAll Test. This result suggests high instructional validity of the OverAll Test.

Table 1. The Degree of Overlap Between the Formal and the Operational Curriculum

Variables	Mean	Standard Deviation	n
RT1*	87%	17,33	34
NRT1	12%	15,67	34
RT2	77,4%	12,64	45
NRT2	22,6%	15,67	45
Definition	22,1%(student) 20,6%(staff)	21,24	45
Comprehension	47% (student) 40,4% (staff)	22,64	45
Analysis	30,9% (student) 39% (staff)	16,58	45

RT: Recognized Topics (1= TOC1, 2= TOC2)
NRT: Not Recognized Topics

Additionally for TOC2, the more topics students indicate as "received much of attention during the meetings", the higher their OverAll Test score (r= .40*). On the other hand, the more topics students indicate as "received moderate attention during the meetings", the lower the OverAll Test scores are (r= -.32*). Probably, students acquired partial knowledge by the informal exchanges they had about the topic. This partial knowledge might impede instead of enhance successful problem analysis. There was only a very weak correlation between topics which received not much attention and the test scores (r= .01).

3.1.3 Conclusion

The instructional validity study suggests there is an important degree of overlap between the formal and operational curriculum, in terms of core concepts as well as in terms of the level of comprehension. Although the students followed different learning paths during the instructional period, students perceived they addressed the core concepts of the formal curriculum on the expected level of comprehension. It can be expected that in this case, with the assessment blueprint deduced from the curricular blueprint, the assessment reflects instruction. The results of the validity study confirm this hypothesis.

However, informal discussions with staff and students have raised doubts about the match between instruction and assessment. It was expected that the OverAll test, as a mode of assessment in alignment with the features of the learning environment, would have a positive effect on learning and teaching. This resulted in the study of the consequential validity of the OverAll Test.

3.2 The Consequential Validity of the OverAll Test

The growing implementation of new modes of assessment has been influenced by the high expectations teachers had about their positive effects. Ramsden (1988, p. 24) argued, "Perhaps the most significant single influence on students' learning is their perception of assessment". It was expected that a change of the assessment from a constructivist perspective would enhance students' learning. Within educational measurement, the shift in conceptualisations of learning and assessment has influenced the developments in the philosophy of validity, as Moss (1992) describes. With the changing conceptions of validity in educational measurement, there is a growing interest paid to the multidimensionality of the concept of validity and the relevance of those dimensions for new modes of assessment (Moss, 1992). Linn, Baker and Dunbar (1991) describe the relevance of the consequences of assessment. They define those consequences as "the intended and unintended effects of assessment on the ways teachers and students spend their time and think about the goals of education." (p. 17). However, as Thomson and Falchikov explain (1998), one element in the assessment research that has received less attention than others is that of the students' perceptions of it. Only a few studies addressed this issue (e.g. Gibbs, 1999, Sambell, McDowell & Brown, 1997; Thomson & Falchikov, 1998). The results of these studies as regards new modes of assessment are conclusive. Gibbs (1999) showed in a series of case studies that "students are tuned in to an extraordinary extent to the demands of the assessment system and even subtle changes to methods and tasks can produce changes in the quantity and in the nature of learning outcomes out of all proportion to the scale of change in assessment."(p. 52). Based on a series of interviews with students experiencing new modes of assessment, the research of Sambell, et al (1997) revealed the impacts of assessment on learning. "Their perceptions of poor learning, lack of control, arbitrary and irrelevant tasks in relation to traditional assessment contrasted sharply with perceptions of high quality learning, active student participation, feedback opportunities and meaningful tasks in relation to alternative assessment" (p. 365).

In both studies, there is not much information on the learning environment of which the assessment is part. One can hypothesize that, in the case of new modes of assessment, an alignment between instruction and assessment will enhance the positive effects of assessment on learning. Additionally, the studies describe perceptions with assessment methods that are relatively new for students and staff. Sambell et al. (1997) indicate that for 5 of the 13 cases, the assessment approach was being used for the first time. In only 4 of the 13 cases, the assessment method was considered typical for that course. Probably the observed effects of the implementation of new modes of assessment are partly due to the innovative character of these assessments.

The case described in this chapter, differs from the previous studies in two aspects. First, at the time of the study, the OverAll Test had been implemented for 8 years. Students and teachers are very familiar with its characteristics and purposes. Second, it is developed according to the features of the problem-based learning environment of which it is part. The instructional validity presented earlier, indicated an acceptable alignment between the formal and operational curriculum and the OverAll Test. It was expected that the OverAll test would enhance learning as a construction of knowledge in order to identify, analyse, solve and evaluate novel, authentic problems.

3.2.1 Research Method

3.2.1.1 Procedure

An evaluation questionnaire was developed to measure students' perception of different quality aspects of the learning environment. The questionnaire items were based on interviews with staff members asking for their expectations about the students' study processes. For all the items in the questionnaire, they expected average scores on the 5-point Likert scale higher than 3.5 and with low standard deviations (<0.5). As a measure of reliability, i.e. internal consistency, the Cronbach Alfa coefficient was calculated. In the different surveys, the coefficient varies between .50 and .63. This indicates a moderate reliability of the instrument.

In order to complement the student surveys and in order to gain a deeper insight in students' perceptions of the OverAll Test, semi-structured interviews were held with groups of students. For reasons of between-respondent triangulation, we additionally interviewed a group of teachers. As Sambell et al (1997, p. 355) indicate: "The act of soliciting the varying perspectives of the range of people involved in the assessment process was crucial in building up a rich, fully contextualized picture of the phenomenon

– alternative assessment – under investigation." The aim was to encourage the students and the teachers to talk freely and openly about their experiences with the OverAll Test, with interviewers providing initial stimuli, using an interview schedule (Powney & Watts, 1987). The semi-structured interview schedules of Sambell et al (1997) were adapted to the Maastricht situation and used for the student interviews as well as for the teacher interviews. Contrary to the design of former questionnaires and interview schedules measuring the effect of conventional assessment on learning, the Sambell et al interview schedules were developed to explore student perceptions of the consequential validity of new modes of assessment. The interviews focused on "what students understood to be required, how they were going about the assessment task(s), and what kind of learning they believed was taking place" (Sambell et al, 1997, p. 355). These questions were the core of the OverAll test study presented here.

Based on the rationale of the Focus Group method, the interviews were conducted in a group session and not in a one-on-one session. The participants in the student groups and in the teacher group were asked to discuss their perceptions of different aspects of the OverAll Test. By asking the stakeholders to discuss these issues with peers, the Focus Group Method intends to generate richer information than it is the case with individual interviews (Churchill, 1996).

The kind of data obtained included students' and staff members' detailed descriptions of how they perceived the OverAll Test. In addition, the data reflect the students' and staff's more general reflections and ideas about assessment.

The interview data for analysis were grouped in themes and structures. The validity of the interpretations rests upon careful reflection and discussions with researchers not involved in the interviews. Therefore, we discussed the interpretations with a team of 3 educational scientists and one economist. Additionally, we searched for confirmatory and contrary evidence in order to strengthen the development of interpretations by a round table discussion with a random sample of the interviewed staff members and students.

3.2.1.2 Sample

For the survey, all students administering the test were asked to answer the questionnaire. The response rate varied between 50% and 70%. For the Focus Group, the interviews were conducted with 5 student groups and a staff member group (n=8). Within a random sample of 5 tutorial groups (n = 12), students were asked to volunteer. In total 48 students participated. Teachers who were engaged in the OverAll test for many years were asked to cooperate. In total 8 staff members participated.

3.2.2 Results

3.2.2.1 The Student Survey

The results revealed some notable phenomena. Table 1 summarizes the average scores of the first year students (n=100, academic year 1998-1999) on a set of items from the questionnaire (5-points Likert scale). The data are consistent over the past years.

Table 2. The Average Scores of Students on the OverAll Test Student's Perceptions Questionnaire (OverAll Test 1, 1998-1999, Faculty of Economics and Business Administration)

	M (s)
Self-study	
• I read the cases	4.5 (1.0)
• I checked the concepts I did not understand	4.1 (1.0)
• I indicated the key concepts in the cases	3.9 (1.1)
• I made schemas of the content of the cases	2.8 (1.2)
• I checked the topics of the cases in the learning materials of the module	3.7 (1.1)
• I tried to formulate figures and tables in the case in my own words	3.6 (0.8)
• When choices or decisions were made in the case, I tried to find arguments from the learning materials of the module	3.2 (0.9)
• I tried to explain the data presented in the case on the basis of the learning materials of the module	3.8 (0.7)
• I tried to understand the cases by asking why-questions	3.4 (1.0)
• I made notes from the analyses I made	3.3 (1.1)
• I worked on the cases together with peers	2.6 (1.7)
• How many hours did you spend working on the cases?	27.0 (16.1)
The OverAll Test	
• The way of working in the tutorial group fits the way of questioning in the OverAll Test	2.5 (1.0)
• The test questions were more difficult than expected	4.1 (0.8)
• For most questions I had to read the cases	2.8 (1.0)
• I had enough time to answer the questions	2.9 (1.4)

It is clear from the results in Table 1 that only reading the cases is a common activity of the students. This finding was quite disappointing for the staff. For many students, analysing in depth the cases was not part of their working activity on the cases. Although students had 2 weeks free of tutorial meetings in order to work on the cases, they spent less than half that time in

Evaluating the OverAll Test 133

this activity. During the early years, the staff tried to motivate and guide the students more by giving more concrete study guidelines together with the cases. This did not change the results of the questionnaire significantly. It was especially the answer to the question "The way of working in the tutorial group fits the way of questioning in the OverAll Test" which surprised the staff. Although a former validity study suggested instructional validity (see above), students did not seem to perceive a match between the processes in the tutorial group and the manner of questioning in the OverAll Test. Particularly because working on problems is the main process within problem based learning environments, the staff considered this students' perception as a serious issue.

3.2.2.2 The Semi-Structured Interviews

Different issues were addressed. First, the students and the staff expressed their views on the concept of the OverAll Test and its relationship with other assessment instruments. Second, views on the relation between instructional practice and assessment practices were expressed. Third, views on the way of working during the self-study period were considered. Fourth, views on how the assessment practices can be optimised were explored.

3.2.2.2.1 The Concept of the OverAll Test and the Knowledge Test

The students and the staff described the OverAll Test with two characteristics: the level of competence measured and the domain questioned.

The students explained the goal of the OverAll Test as measuring the application of knowledge. As Sebastian said: "in the OverAll Test you have to use knowledge in practice". Thomas explained it as follows: "The OverAll Test asks you to use knowledge; you need to do more than for the Knowledge Test. For the Knowledge Test you read the textbooks and study it by heart. For the OverAll Test, you have to relate things; you have to cope with the context where knowledge is to be used. The OverAll Test is building knowledge, the Knowledge Test is memorising." Stephanie, a tutor, used the term "the linking of knowledge to practice."

Concerning the domain characteristic, the students perceived the OverAll Test as asking the students to connect theories. Tobias said: "The OverAll Test is a summary of two blocks (instructional periods). It checks if you remembered the basics of two blocks." René, a tutor, described the OverAll Test as a kind of test, where you progressively measure to what extent students are able to use knowledge from different instructional periods."

3.2.2.2.2 Match between Instruction and Assessment

The staff as well as the students indicated that, in theory, the transfer of the problem-solving skills from the tutorial group to the self-study period and the OverAll Test should be "a natural step", as Peter said. However, they experienced the tutorial group as too much stressing the reproduction of what was read in the literature. The literature was supposed to be a tool for explaining, analysing and solving the problem posed. In practice, the problem was often used only as a starting point for going to literature. From that moment on, the literature became a goal instead of a tool. The students experienced this process as highly influenced by the skills of the tutor. The staff formulated three reasons for the "reproductive" functioning of some tutorial groups: the skills of the tutor, the amount of concepts that are addressed in the instructional period, and the motivation of the students. Mark, a staff member said: "It is the task of the tutors to help the students to understand the context of what they learn. The motivation of the student influences the extent to which this happens. At the time, there is too much reproduction of knowledge in the tutorial group and too little creativity. This can also be seen in the students' answers on the OverAll Test items. They reproduce knowledge that is relevant for the problem analysis questions, but they do not link the theory to the case. But, can we require creativity from the students if we did not pursue it during the tutorials?" Maarten, a colleague added: "Too many subjects are planned within the curriculum. There is no time left for discussion and for going back to the problem."

The students asked for more exercises and discussion. Kurt, a student, stated this point very clear: "How should you be able to analyse things if you have to deal with 19 chapters within 6 weeks? There the problem-based learning system fails". He added, "within the tutorials, the graphs were less complex and mostly they were drawn in the book and you had to interpret them. In the OverAll Test, you had to draw the graphs yourself." The students also indicated that they had problems with the novelty of the problems. During the tutorials, the learning took place based on a problem. Discussions of the key issues based on similar problems or problems with slight variations on the starting problem, seldom took place. "During the OverAll Test, you suddenly have to deal with novel problems. Sometimes those problems questioned in the OverAll Test present variations which are difficult to work with," the students concluded.

Another aspect discussed was the feedback on what students know. Because some of the tutorial groups only end up with summarizing the literature without discussing in depth the relevance to the problem, there was no real feedback on student's understanding of the key issues. This feedback occurs when the students start discussing what they found in literature and its relation to the problem as presented in the case. Improving the tutoria

Evaluating the OverAll Test 135

means improving this feedback function, according to the students and the staff. Additionally, both expressed the need for feedback on the test results. Then, the real learning starts.

In order to improve feedback and to get the first year students acquainted with the OverAll Test, in the middle of the first block, students received a novel case tackling issues studied during the past three weeks based on different problem tasks. The students were asked to answer a set of problem analysis questions based on this case, which were similar to OverAll Test items. The answers to these questions as well as the problem-analysis process were discussed within the tutorial group. The students felt this was relevant to do and they encouraged this way of giving feedback and exercising. However, because the OverAll Test was at the end of the next six weeks block, they perceived it as "not primarily relevant for the moment".

These perceptions suggest that because of curricular as well as tutor problems, the tutorial groups did not sufficiently succeed in relating knowledge to practice. This is mirrored in the problems students face when analysing the cases that are subject of the OverAll Test. Additionally, the feedback function of the tutorial group failed to a certain extent. The feedback function was overwhelmed by the reproduction of a large amount of content matters relevant to the starting problem. Moreover, the explicit feedback moment during the block was experienced as only one single moment of exercising, inappropriately planned in the curriculum. The students expressed the need to do this exercise as flexible problem analysis as part of all blocks.

3.2.2.2.3 Students' Activities

The staff expressed the feeling that the students do not work enough during the self-study period. In some cases, the students only read the articles. The students agreed that for some of them, reading the cases, sometimes, was the only activity. Other students formed a small tutorial group themselves and discussed the cases. The students of the group indicated they experienced this process as very effective. "It drives you to think critically about what you found" David said.

The students mentioned different reasons for not working full-time during the self-study period. Sometimes, they experienced the cases as not interesting. "Sometimes, especially when the cases are long, you do not know what to do with it ", Dirk said. "For the OverAll Test, you need to do more. But what this "doing more" exactly means, is not clear." Some students expressed that they did not know how to start working. If they read the cases and checked the relevant issues in the literature, what should they do more?

This feeling of being unsure how to handle the cases mirrors the problems of the group functioning mentioned earlier. As in some tutorial groups, the process sticks after the content matters are reproduced. The going back from theory to practice in order to better understand practice, was missing.

Finally, the students referred to the minor weight of the OverAll Test in the final examinations. If it had more weight, they surely would put more energy into it.

3.2.2.2.2.4 Optimising Assessment Practices

The concept and the relevance of the OverAll Test are largely accepted. The students as well as the staff members indicated the OverAll Test is an inherent aspect of the problem-based learning curriculum. It is instructional practice that fails to some degree.

According to the problems expressed, students as well as staff members asked for more feedback, more time for discussion and for using knowledge as a tool for problem-solving, and for more skilled tutors.

4. CONCLUSION

As the core goals and features of the learning environment changed during the last decade, questions were posed as to what extent and in what sense assessment of students' performances should be adapted to these new directions in learning and instruction. This has led to the expanding implementation of new modes of assessment. The development of the OverAll Test in a problem-based learning environment is an example of these changes in instructional and assessment approach.

The shift in conceptions of learning, teaching and assessment, together with the expansion of the implementation of new modes of assessment, has led to shifting conceptions of validity in educational measurement. Although many debates have been going on, only a few research studies address the quality of new modes of assessment from this new perspective.

The present study indicates the value-added of looking for evidence of multiple dimensions of validity of new modes of assessment. Concerning the match between instruction and assessment, it seemed that the OverAll Test measured the concepts at the level of comprehension, perceived by the students as part of instruction. However, one important concern was the intended and unintended effects of the OverAll test on the way students and teachers spend their time and think about the goals of education and, in particular, assessment. The results of a yearly student survey and of the semi-structured interviews with groups of students and teachers have led to

Evaluating the OverAll Test 137

the following conclusions. Students as well as teachers agreed largely on what they understood to be required by the OverAll Test: students have to apply knowledge; they have to cope with the context where knowledge is to be used. The OverAll Test is about building knowledge. They both recognised that the OverAll Test requires accessing previously acquired knowledge in a new set of contexts. Taking into account these characteristics, students as well as the staff perceived the OverAll test as relevant, as an inherent aspect of the problem-based learning environment. Concerning how the students were going about the assessment tasks, it seemed that the students spend less than half of the time planned for working on the novel cases. The students indicated that to a large extent, they did not get further than reading the case and reading about the relevant concepts in the literature. This was mirrored in the way the students and the teacher perceive the learning that is taking place in the tutorial groups. In some cases, the tutorial groups ended up summarizing the theory that was relevant for the starting problem. There was no in-depth analysis of the starting problem and no transfer of the general concepts derived from this starting problem to similar problems. For some tutorial groups, using the theory as a tool for analysing and solving the problem was not the core practice. Students as well as staff referred to the overloading of the blocks and the problem of some tutors who were not skilled to stimulate the group to discuss, analyse and use the knowledge as a tool for solving a variety of problems. This too "reproductive" functioning of the tutorial group was a burden to achieve in-depth feedback on the learning process.

4.1 Recommendations for Assessment and Instruction

The findings reported indicate the importance of the alignment of instruction and assessment. In this respect, Biggs (1996) stresses the importance of constructive alignment, where students' performances that represent a high cognitive level (understanding), are nominated in the objectives and thus used to systematically align teaching methods and the assessment. Under these conditions, new modes of assessment such as case-based assessment and the OverAll Test can influence student learning. However, the subjective learning environment, the way students perceive the learning environment, seems to play a mediating role. Although these new modes of assessment encourage study behaviour such as "building knowledge", the quality of the learning environment as perceived by the students plays a crucial role in the extent to which students really engage in these kind of study behaviours. The research results presented here indicate that, when interpreting the results of consequential validity studies of new modes of assessment, the perceived learning environment has to be taken into account.

It can be concluded that the evaluation of the assessment practices lead to recommendations for improving instruction. It seems that, as students express, the burden is on the instructional practice and not on the assessment instrument.

4.2 Recommendations for Future Research

It is commonly stressed that "the tail wags the dog": the assessment influences to a large extent what and how students learn. However, the findings on the consequential validity of the OverAll test, presented in this chapter, indicate that the relation between assessment and learning is more complex. The learning environment as perceived by the students, can mediate the effect of the assessment practices on learning and teaching. Researchers investigating the consequential validity of new modes of assessment should take this into account.

In edumetrics, validity is stressed as a crucial quality indicator for new modes of assessment. The consequential validity study presented here is conducted within this edumetric framework. However, additional research is necessary to measure the various quality aspects of the OverAll Test. Four additional research questions can be formulated.

Is the OverAll Test fair to all the students? Especially with respect to performance assessments, this question is often raised (Bond, 1995). Within the case presented, students from different nationalities with different learning experiences and different learning styles are working in small groups. To what extent is the OverAll Test fair for different subpopulations with respect to their nationality, prior experiences, learning styles and the tutorial group attended?

The notion of predictive validity asks for a procedure of determining the extent to which this assessment instrument predicts accurately the performance of the students in their subsequent study careers (Benett, 1993) This issue can be operationalized as: do high performers on the OverAll Test perform better on the projects they do in graduate courses, and is there an effect on their entrance on the labour market?

Finally, the question remains of whether the results of the studies reported are case-specific. Research in other settings, with other curricula and with other student populations where the OAT is implemented, can indicate how generalizable are the results obtained in this study. Comparing the consequential validity of the OverAll test in various curricula and learning environments can indicate under which conditions (i.e. learning environments) the OverAll test optimally stimulates learning as a construction of knowledge in order to identify, analyse, solve and evaluate novel, authentic problems.

REFERENCES

Barrows, H. S. (1986). A taxonomy of problem-based learning methods. *Medical Education, 20*, 481-186.
Benett, Y. (1993). The validity and Reliability of Assessments and Self-assessments of Work-based Learning. *Assessment and Evaluation in Education, 18* (2), 81-94.
Biggs, J. (1996). Enhancing teaching through constructive alignment. *Higher Education, 32*, 347-364.
Birenbaum, M. (1996). Assessment 2000: Towards a Pluralistic Approach to Assessment. In M. Birenbaum, & F. J. R. C. Dochy (Eds.), *Alternatives in Assessment of Achievements, Learning Processes and Prior Knowledge* (pp. 3-29). Boston: Kluwer Academic Press.
Birenbaum, M., & Dochy, F. (Eds.). (1996). *Alternatives in Assessment of Achievement, Learning Processes and prior Knowledge*. Boston: Kluwer Academic.
Bond, L. (1995). Unintended Consequences of Performance Assessment: Issues of Bias and Fairness. *Educational Measurement: Issues and Practice, Winter 1995*, 21-24.
Bowden, J., & Marton, F. (1998). *The University of Learning*. London: Kogan Page
Brown, S., & Knight, P. (1994). *Assessing learners in higher education*. London: Kogan Page.
Calfee, R. (1983). Establishing instructional validity for minimum competence programs. In G. F. Madaus, *The courts, validity, and minimum competence testing* (pp. 95-114). Boston: Kluwer-Nijhoff Publishing.
Churchill, G. A. (1996). *Basic Marketing Research*. Orlando: The Dryden Press.
De Corte, E. (1990). *A State-of-the-art of research on learning and teaching*. Keynote lecture presented at the first European Conference on the First Year Experience in Higher Education, Aalborg University, Denmark, 23-25 April 1990.
De Haan, D. M. (1992). *Measuring test-curriculum overlap*. Enschede: Febo.
Des Marchais, J. E., & Dumais, B. (1993). An Attempt at Measuring Student Ability to Analyze Problems in the Sherbrooke Problem-Based Curriculum: a Preliminary Study. In D. Boud & G. Feletti (Eds.), *The challenge of problem based learning*. London: Kogan Page.
Dolmans, D. (1994). How students learn in a problem-based curriculum. Maastricht: Universitaire Pers.
Engel, C. E. (1991). Not just a method but a way of learning. In D. Boud & G. Feletti (Eds.), *The challenge of problem based learning*. London: Kogan Page.
English, F. W. (1992). *Deciding what to teach and test*. Newbury Park California: Sage Publications Company, Corwin Press, INC.
Feltovich, P. J., Spiro, R. J., & Coulson, R. L. (1993). Learning, Teaching, and Testing for Complex Conceptual Understanding. In N. Frederiksen, R. J. Mislevy, & I. I. Bejar (Eds.), *Test theory for a New Generation of Tests* (pp. 178-193). Hillsdale, New Jersey: Lawrence Erlbaum Associates, Publishers.
Gibbs, G. (1999). Using assessment strategically to change the way students learn (pp. 40-56). In S. Brown & A. Glasner (Eds.). *Assessment matters in Higher Education*. Milton Keynes: Open University Press.
Glaser, R. (1990). Toward new models for assessment. *International Journal of Educational Research, 14*, 475-483.

Leinhardt, G., & Seewald, A. M. (1981). Overlap: What's Tested, What's Taught? *Journal of Educational Measurement, 18 (2),* 85-95.

Linn, R. L., Baker, E., & Dunbar, S. B. (1991). Complex, performance-based assessment: Expectations and validation criteria. *Educational Researcher, 16,* 1-21.

McClung, M. S. (1979). Competency testing programs: Legal and educational issues. *Fordham Law review, 47,* 6511-712.

Moss, P. A. (1992).Shifting conceptions of validity in Educational Measurement: Implications for Performance Assessment. *Review of Educational Research, 62* (3), 229-258.

Nuy, H. J. P. (1991). Interaction of study orientation and students' appreciation of structure in their educational environment. *Higher Education, 22,* 267-274.

Poikela, E., & Poikela, S. (1997). Conceptions of Learning and Knowledge – Impacts on the Implementation of Problem-based Learning. *Zeitschrift fur Hochschuldidactik, 21* (1), 8-21.

Powney, J., & Watts, J. (1987). *Interviewing in educational research.* London: Routledge Kegan Paul.

Ramsden, P. (1979). Student learning and the perceptions of the academic environment, *Higher Education, 8,* 411-428.

Ramsden, P. (1998). *Improving Learning. New perspectives.* London: Kogan Page.

Sambell, K., McDowell, L., & Brown, S. (1997). "But is it fair?": an exploratory study of student perceptions of the consequential validity of assessment. *Studies in Educational Evaluation, 23* (4), 349-371.

Savery, J. R., & Duffy, T. M. (1995). Problem Based Learning: An Instructional Model and Its Constructivist Framework. *Educational Technology, Sept-Oct.,* 31-38.

Schoemaker, P. J. H. (1995). Scenario Planning: a Tool for Strategic Thinking. *Sloan Management Review,* 25-39.

Segers, M. S. R. (1996). Assessment in a Problem-Based Economics Curriculum. In M. Birenbaum, & F. J. R. C. Dochy, F. (Eds.), *Alternatives in Assessment of Achievements, Learning Processes and Prior Knowledge* (pp. 201-224).Boston/ Dordrecht/ London: Kluwer Academic Publishers.

Shavelson, R. J., Gao, X., & Baxter, G. P. (1996). On the content validity of performance assessments: Centrality of domain specification. In M. Birenbaum, & F. Dochy (Eds.), *Alternatives in Assessment of Achievements, Learning Processes and Prior Learning* (pp. 131-143). Boston: Kluwer Academic Press.

Thomson, K., & Falchikov, N. (1998). "Full on Unitil the Sun Comes Out": the effects of assessment on student appoaches to studying. *Assessment and Evaluation in Higher Education, 23* (4), 379-390.

Assessment for Learning: Reconsidering Portfolios and Research Evidence

Anne Davies[1] & Paul LeMahieu[2]
[1]Classroom Connections International, Canada, [2]University of California, Berkeley, USA

Only if we expand and reformulate our view of what counts as human intellect will we be able to devise more appropriate ways of assessing it and more effective ways of educating it.
Howard Gardner

1. INTRODUCTION

Since the late 1980's, public education in North America has been shifting to a standards or outcomes based and performance oriented systems. Within such systems, the most basic purpose of all education is student learning, and the primary purpose of all assessment is to support that learning in some fashion. The assessment reform that began in the 1980's in North America has had numerous impacts. Most significantly, it has changed the way educators think about students' capabilities, the nature of learning, the nature of quality in learning, as well as what can serve as evidence of learning in terms of classroom assessment, teacher assessment and large-scale assessment. In this context, the use of portfolios as a mode of assessment has gained a lot of interest. This chapter will explore the role of assessment in learning and the role portfolios might play. Research evidence of the qualities of portfolios for enhancing student learning is presented and discussed.

2. THE ROLE OF ASSESSMENT IN LEARNING

Learning occurs when students are, "thinking, problem-solving, constructing, transforming, investigating, creating, analysing, making choices, organising, deciding, explaining, talking and communicating, sharing, representing, predicting, interpreting, assessing, reflecting, taking responsibility, exploring, asking, answering, recording, gaining new knowledge, and applying that knowledge to new situations." (Cameron, Tate, Macnaughton, & Politano, 1998, p 6). The primary purpose of student assessment is to support this learning. Learning is not possible without thoughtful use of quality assessment information by learners. This is reflected in Dewey's (1933) "learning loop," Lewin's (1952) "reflective spiral," Schön's (1983) "reflective practitioner," Senge's (1990) "reflective feedback," and Wiggin's (1993) "feedback loop." Education (K-12 and higher education) tends to hold both students and teachers responsible for learning. Yet, if students are to learn and develop into life long, independent, self-directed learners they need to be included in the assessment process so the "learning loop" is complete. Reflection and assessment are essential for learning. In this respect, the concept of assessment for learning as opposed to assessment of learning, has emerged.

For optimum learning to occur students need to be involved in the classroom assessment process. When students are involved in the assessment process they are motivated to learn. This appears to be connected to choice and the resulting ownership. When students are involved in the assessment process they learn how to think about their learning and how to self-assess – key aspects of meta-cognition. Learners construct their own understandings therefore, learning how to learn - becoming an independent, self-directed, life long learner - involves learning how to assess and learning to use assessment information and insights to adjust learning behaviours and improve performance behaviours and improve performance.

Students' sense of quality in performance and expectations of their own performance are increased as a result of their engagement in the assessment process. When students are involved in their learning and assessment they have opportunities to share their learning with others whose opinions they care about. An audience gives purpose and creates a sense of responsibility for the learning which increases the authenticity of the task (Davies, Cameron, Politano, & Gregory, 1992; Gregory, Cameron, & Davies, 2001; Sizer, 1996).

Students can create more comprehensive collections of evidence to demonstrate their learning because they know and can represent what they've learned in various ways to serve various purposes. This involves gathering evidence of learning from a variety of sources over time and looking for patterns and trends.

The validity and reliability of classroom assessment is increased when students are involved in collecting evidence of learning. The collections are more likely to be more complete and comprehensive than if teachers alone collect evidence of learning. Additionally, this increases the potential for instructionally relevant insights into learning. Teachers employ a range of methods to collect evidence of student learning over time. When evidence is collected from three different sources over time, trends and patterns can become apparent. This process has a history of use in the social sciences and is called triangulation (Lincoln & Guba, 1984). As students learn, evidence of learning is created. One source of evidence are products such as tests, assignments, students' writings, projects, notebooks, constructions, images, demonstrations, as well as photographs, video, and audiotapes. They offer evidence of students' performances of various kinds across various subject areas. Observing the process of learning includes observation notes regarding hands-on, minds-on learning activities as well as learning journals. Talking with students about their learning includes conferences, written self-assessments, and interviews. Collecting products, observing the process of learning, and talking with students provides a considerable range of evidence over time

Taking these critical success factors of learning into account, portfolio as a mode of assessment poses unique challenges.

3. PORTFOLIO AND ITS CHARACTERISTICS

Gillespie, Ford, Gillespie, & Leavell, (1996) offers the following definition: "Portfolio assessment is a purposeful, multidimensional process of collecting evidence that illustrates a student's accomplishments, efforts, and progress (utilising a variety of authentic evidence) over time." (p. 487). In fact, portfolios are so purposive that everything that defines a portfolio system:
1. what is collected;
2. who collects it;
3. how it is collected;
4. who looks at it;
5. how they look at it; and
6. what they do with what they see.
are all determined first by the purpose for the portfolio.

Consider, for example, a portfolio with which one will seek employment. While there must be no duplicity in the evidence presented, it would seem perfectly acceptable, even expected, that the candidates will present themselves in the best possible light. What is most likely to find its way into such a portfolio is a finished product – and often the best product at that. On the other hand, consider a portfolio with which a student reveals to a trusted teacher a completely balanced appraisal of his or her learning: strengths, certainly, but also weaknesses as well as the kinds of processes the learner uses to produce his or her work. This portfolio is likely to have a number of incomplete efforts, some missteps, and some product that reveals current learning needs. This is not the sort of portfolio with which one would be comfortable seeking employment. The point is a simple one: while they appear similar in many respects, portfolios are above all else purposive and everything about them derives from their desired purpose. This is why some frank discussion about purpose at the outset of developing a portfolio system is essential. Often, when teachers feel blocked about some decision about their portfolio process, the answer is apparent upon remembering their purpose.

There is no one or best specific purpose for portfolios. Portfolios can be used to show growth over time (e.g. Elbow, 1986; Politano et al., 1997; Tierney, Carter, & Desai, 1991), to provide assessment information that guides instructional decision-making (e.g., Arter & Spandel, 1992; Gillespie et al., 1996; LeMahieu & Eresh, 1996a), to show progress towards curriculum standards (e.g. Biggs, 1995; Gipps 1994; Frederiksen & Collins, 1989; Sadler, 1989a), to show the journey of learning including process and products over time (e.g. Costa & Kallick, 2000; Gillespie et al., 1996) as well as used to gather quantitative information for the purposes of assessment outside the classroom (e.g. Anson & Brown, 1991; Fritz, 2001; Millman, 1997; Willis, 2000).The strengths of portfolios is that of range and comprehensiveness of evidence, variety and flexibility in addressing purpose (Julius, 2000).

Portfolios are used successfully in different ways in different classrooms. Portfolios are generally defined in the literature in terms of their contents and purpose - an overview of effort, progress or performance in one or several subjects (e.g. Arter & Spandel, 1992; Gillespie et al., 1996; Herman Aschbacher, & Winters, 1992). There are numerous examples of student portfolios developed to show learning to specific audiences in different areas. They are being used in early childhood classes (e.g. Potter, 1999. Smith, 2000), with students who have special needs (e.g. Law & Eckes 1995; Richter, 1997), and in elementary classrooms for Science (e.g. Valdez 2001) for writing (Howard & LeMahieu, 1995; Manning, 2000), and mathematics (Kuhs, 1994). Portfolios in high schools were used initially in performance-based disciplines such as fine arts, then in writing classes, and have now expanded to be used across many disciplines such as science education (e.g. Reese, 1999), academic and beyond (e.g. Konet, 2001), chemistry classes (e.g. Weaver, 1998), English classes (e.g. Gillespie et al., 1996), and music education (e.g. Durth, 2000). There is a growing body of research related to electronic portfolios (e.g. Carney, 2001; Quesada, 2000; Young, 2001; Yancey & Weiser, 1997). Portfolios are also being used in teacher-education programs and in higher education more broadly (e.g. Kinchin, 2001; Klenowski, 2000; McLaughlin & Vogt, 1996; Schonberger, 2000).

There is a range of evidence students can collect. Also, since there are different ways for students to show what they know, the assessment information collected can legitimately differ from student to student (see for example Anthony, Johnson, Mickelson, & Preece, 1991; Gardner & Boix-Mansilla, 1994).Collecting the same information from all students may not be fair and equitable because students show what they know in different ways (e.g. Gardner, 1984; Lazear, 1994). When this assessment information about learning is used to adjust instruction, further learning is supported. Evidence of learning will also vary depending on how students represent their learning. Portfolios uniquely provide for this range of expression of learning. When they are carefully developed, they do so with evidence that can be of considerable technical quality and rigor.

From an assessment perspective, portfolios provide at least four potential "values-added" to more traditional means of generating evidence of learning:

1. they are extensive over time and therefore reveal growth and development over time (however simply or subtly the growth may be defined);
2. they allow for more sustained engagement and therefore permit the examination of sustained effort and deeper performance;
3. to the extent that choice is involved in the selection of content (both teacher and most especially student choice), then portfolios reveal students' understandings about and dispositions towards learning (including the unique special purposes that portfolios might address and their consequent selection guidelines); and
4. they offer the opportunity for students to interact with and reflect upon their own work.

It is important to note that for portfolios to realise their potential as evidentiary bases for instructional decision-making, then particular attention must be given to some one (or all) of these four "values-added." Not only should they serve as the focus for generating evidence uniquely beneficial to portfolios, but care must be taken in the construction and application of evaluative frameworks such that rigor and discipline attends the generation of data relevant to some one or all of these points.

Allowing for a range of evidence encourages students to represent what they know in a variety of ways and gives teachers a way to fairly and more completely assess the learning. Collecting information over time provides a more comprehensive picture. For example, Elbow and Belanoff (1991) stated, "We cannot get a trustworthy picture of a student's writing proficiency unless we look at several samples produced on several days in several modes or genres" (p. 5). Portfolios may contain writing samples, pictures, images, video or audiotapes, work samples – different formats of evidence that helps an identified audience understand the student's accomplishments as a learner.

There are numerous ways students are involved in communicating evidence of learning as presented in portfolios. Some examples. Portfolios complement emerging reporting systems such as student, parent, teacher conferences (Davies et al., 1992; Davies, 2000; Gregory et al., 2001; Macdonald, 1982; Wong-Kam, Kimura, Sumida, Ahuna-Ka`ai, & Hayes Maeshiro, 2001). Sometimes students and parents meet at school or at home to review evidence of learning often organised into portfolios to show growth or learning over time (Davies et al., 1992; Howard & LeMahieu, 1995). Other times portfolios are used in more formal conference settings or exhibitions where students present evidence of learning and answer questions from a panel of community members, parents, and peers (Stiggins & Davies, 1996; Stiggins, 1996, 2001). Exhibitions are part of the graduation requirements in schools belonging to the Coalition of Essential Schools (Sizer, 1996). Sometimes students meet with teachers to present their learning and the conversation is between teacher and student in relation to the course goals (Elbow, 1986). This format appears more appropriate for older high school students and for graduate and under-graduate courses. In a few instances, portfolios have been developed (including student choice in their assembly) for evaluation and public accounting of the performance of a program, a school, or a district. (LeMahieu, Eresh, & Wallace, 1992b; LeMahieu, Gitomer, & Eresh, 1995a). This approach, when defined as an active process of inquiry on the part of a concerned community transforms accountability from a passive enterprise in which the audience is "fed" summary judgements about performance to an active process of coming to inspect evidence and determine personal views about the adequacy of performance and (even more important) recommending how best to improve it (Earl & LeMahieu, 1997a; LeMahieu, 1996b). All of these ways of communicating have one thing in common – the student is either present or actively represented and involved in presenting a range of evidence of learning. The teacher assists by providing information regarding criteria and evidence of quality. Sometimes this is done through using a continuum of development that describes learning over time using samples from large-scale portfolio or work sample assessments. These samples provide a reference point for conversation about student development and achievement. Teachers use samples of work that represent levels of quality to show parents where the student is in relation to the expected standard. This helps respond to the question many parents ask, "How is my child doing compared to the other students?" These kinds of conferences involve parents and community members as participants in understanding the evidence and in "reporting" on the child's strengths, areas needing improvement and the setting of goals. This kind of "verbal report card" involves students, parents, and the teacher in a face-to-face conversation supported with evidence.

4. PORTFOLIOS AND THEIR QUALITIES

4.1 The Reliability and Validity Issue

When portfolios are used for large-scale assessment, concerns around their reliability and validity are expressed. For example, Benoit and Yang (1996), after using portfolios for assessment at the district level, recommend clear uniform content selection and judgement guidelines because of the need for greater inter-rater reliability and validity. Berryman and Russell (2001) indicates a similar concern for ensuring reliability and validity when he reports the Kentucky statewide rate for scoring the portfolios is 75% for exact agreement their school scoring has "86% exact agreement." Resnick and Resnick, (1993) reported that while teachers refined rubrics and received training, it was a challenge to obtain reliability between scorers. Inter-rater reliability of portfolio work samples continues to be a concern (e.g. Chan, 2000; Fritz, 2001; Willis, 2000). Fritz (2001; p. 32) "The evaluation and classification of results is not simply a matter of right and wrong answers, but of inter-rater reliability, of levels of skill and ability in a myriad of areas as evidenced by text quality and scored by different people, a difficult task at best." Clear criteria and anchor papers assist the process. Experience seems to improve inter-rater reliability (Broad, 1994; Condon & Hamp-Lyons, 1994; White, 1995; 1994b).

DeVoge (2000), whose dissertation examined the measurement quality of portfolios, notes that standardisation of product and process led to acceptable levels of inter-rater reliability. Concerns regarding portfolios being used for gathering information across classrooms within schools, districts, and provinces/states are expressed particularly in regard to large scale portfolio assessment projects such as Kentucky and Vermont's statewide portfolio programs and Pittsburgh Public Schools (LeMahieu, Gitomer, & Eresch, 1995a). Some researchers express concerns regarding reliability (e.g. Calfee & Freedman, 1997; Callahan, 1995; Gearhart, Herman, Baker, & Whittaker, 1993; Koretz, Stecher, & Deibert, 1993; Tierney et al., 1991) while others point out the absence of certain controls as would give confidence even as to whose work is represented in the collected portfolio (Baron, 1983). Novak, Norman and Gearhart (1996) note that the difficulties stem from "variations among the project portfolios models, models that differ in their specifications for contents, for rubrics, and for methods for applying the rubrics." (p. 6) Novak et al. (1996) examined techniques for assessing student writing. Raters were asked to score collections of elementary student narratives using holistic scales from two rubrics. Comparisons were based on three methods and results were mixed. One rubric gave good evidence of reliability and developmental validity. They sum up by noting that "if appropriate cut points are set then reasonably consistent decisions can be made regarding the mastery/non-mastery of the narrative writing competency of third grade students using any rubric-assessment combinations with one exception" (p. 30).

Fritz (2001) names eight studies where researchers are seeing continued improvement in the quality of the data from portfolio assessments (p. 28). Fritz (2001) studied the level of involvement in the Vermont Mathematics Portfolio assessment in Grade 4 classrooms. In particular she was interested in whether involvement in the scoring process let to improved mathematics instruction. She explains that the student portfolio system requires a stratified random sample of mathematics that is centrally scored using rubrics developed in Vermont. The portfolio pieces are submitted on alternate years. In 1996 87% of schools that have Grade 4 students submitted mathematics portfolios. In 1996 teachers at 91 of 350 schools scored mathematics portfolios. She notes that the Vermont procedures have been closely examined with a view to improving the scoring (Koretx, Stecher, Klein, McCaffrey & Deibert, 1993). Current procedures are similar to those used in the New Standards Project (Resnick & Resnick, 1993). Individual teachers score student work. Up to 15% of the papers are double scored and those papers are compared to each other to check for consistency between scorers.

Researchers reporting good levels of reliability in scoring performance assessments include Arter, Spandel and Culham, 1995; Gearhart, Herman, & Novak; 1994; LeMahieu, Gitomer, & Eresh; 1995. Herman (1996) summarises the issues relating to validity and reliability. She explains that while challenging, "assuring the reliability of scoring is an area of relative technical strength in performance assessment" (p. 13). Raters can be trained to score open-ended responses consistently. For example, Herman (1996) reports the Iowa Tests of Basic Skills direct writing assessment demonstrates it is possible to achieve high levels of agreement with highly trained professional raters and tightly controlled scoring conditions (Herman cites Hoover & Bray, 1995). She goes on to note that portfolio collections are more complex and this multiplies the difficulty of ensuring reliability. LeMahieu, Gitomer and Eresh report reliabilities ranging from .75 to .87 and inter-rater agreement rates ranging from 87% to 98% for a portfolio system developed in a large urban school district. They go on to document the steps taken to ensure these levels of reliability (LeMahieu, Gitomer, & Eresh, 1995a, 1995b). These included involving teachers in the inductive process that developed and refined the assessment frameworks (including rubrics) and drawing upon such development partners as model scorers; extensive training for all scorers (development partners as well as new scorers) that includes observation of critical reviews of student work by model scorers, training to an acceptable level of criterion performance for all scorers, using benchmark portfolios that are carefully selected as part of the development process to illustrate both the nature of performance at various levels as well as some of the more common issues in the appraisal of student work; constant accommodation processes during the scoring with adjudication of discrepant score as needed.

Despite the positive research results concerning inter-rater reliability, Darling-Hammond (1997) after reviewing information regarding portfolio and work sampling large-scale assessment systems questioned whether they resulted in improvements in teaching and learning as well as whether or not they were able to measure quality of schooling. In this sense, Darling-Hammond, in line with the expanded view on validity in edumetrics, asks for more evidence for the consequential validity of portfolio assessment. To what extent is there evidence that portfolio assessment leads to the theoretically assumed benefits for learning?

4.2 Do Portfolios Lead to Better Learning and Teaching?

Student portfolios are usually promoted as a powerful instrument for formative assessment or for assessment for learning (e.g. Elbow & Belanoff, 1986; Julius, 2000; Tierney et al., 1991). Portfolios are viewed as having the potential to allow learners (of all ages and kind) to show the breadth and depth of their learning (e.g. Berryman & Russell, 2001; Costa & Kallick, 1995; Davies, 2000; Flood & Lapp 1989; Hansen, 1992; Howard & LeMahieu, 1995; Walters, Seidel & Gardner, 1994; Wolf, Bixby, Glenn, & Gardner, 1991). Involving students in every part of the portfolio process is critical to its success as a learning and assessment tool. Choice and ownership, opportunities to select evidence and reflect what it illustrates while preparing evidence for an audience whose opinion they care about are key aspects of portfolio use in classrooms. Giving students choices about what to focus on next in their learning, opportunities to consider how to provide evidence of their learning (to show what they know), and to reflect and record the learning the evidence represents makes it more possible to learn successfully. Research examining the impact of the use of portfolio's on students' learning focuses on the impact of portfolios on learning in terms of students' motivation, ownership and responsibility, feedback, and self reflection.

4.3 Portfolios: Inviting Choice, Ownership and Responsibility

When learners are engaged, they are more likely to learn. Researchers studying the role of emotions and the brain say that learning experiences such as these prepare learners to take the risks necessary for learning (Goleman, 1995; Jensen, 1998; Le Doux, 1996). Portfolios impact positively on learning in terms of increased student motivation, ownership, and responsibility (e.g. Elbow & Belanoff, 1991; Howard & LeMahieu, 1995; Paulson, Paulson, & Meyer, 1991). For example, Howard and LeMahieu (1995) report that when students in a classroom environment kept a writing portfolio during the school year and shared that portfolio with parents, the students' commitment to writing increased and their writing improved. Researchers studying the role of motivation and confidence on learning and assessment agree that student choice is key to ensuring high levels of motivation (Covington, 1998; Stiggins, 1996). When students make choices about their learning, motivation and achievement increases, when choice is absent, they decrease (DeCharms, 1968; 1972; Deci & Ryan, 1985; Jensen 1998; Lepper & Greene, 1975; Maehr, 1974; Mager & McCann, 1963 Mahoney, 1974; Purkey & Novak, 1984; Tanner, 2000; Tjosvold, 1977 Tjosvold & Santamaria, 1977). Researchers studying portfolios found tha when students choose work samples the result is a deeper understanding o content, a clearer focus, better understanding of quality product, and ar ownership towards the work that "... created a caring and an effort no present in other learning processes" (Gearhardt & Wolf, 1995, p. 69) Gearhart and Wolf (1995) visited classroom at each of four school sites jus before or just after students made their choices for their portfolios. The talked extensively with teachers and students, and collected copies o portfolios. Their project was designed to clarify questions about issues in th implementation of a portfolios assessment program. They noted tha students' choices influenced the focus of personal study, ongoin discussions and the work of the classroom. The increased learning within those classes seemed to be related to students' active engagement through choice. There was also a change in the student/instructor relationship which they report this relationship became more focused, less judgmental and more productive. They note that a balance is needed between external criteria used by the teacher and the internal criteria of the students and conclude by encouraging an on-going dialogue concerning assessment and curriculum amongst students, teachers, and assessment experts.

Tanner (2000) examined the experience with writing portfolios in general education courses at Washington State University. Specifically, he examined issues such as history of the portfolio efforts, experience in light of research, impact on students. Since 1986 students have been required to submit a portfolio that includes three previously produced papers as well as a timed written exam. Later in their studies there is a requirement for a senior portfolio to be determined by the individual disciplines and to be evaluated by faculty from those same disciplines. Tanner notes that the literature references learning through portfolio use in terms of, "student attitude, choice, ownership, performance based learning, growth in tacit knowledge, and the idea of a climate of writing" (p. 83) In his conclusions he affirms that these same elements are present as a result of the portfolio work at Washington State University. Tanner (2000) writes, "... such personal investment, and ownership are the first steps in dialectic participation where ideas and knowledge are owned and remembered, a classic definition of learning." (p. 59) "K-12 research shows connections between learning and such elements as choice and personal ownership of work, elements fostered by portfolio requirements. The connections between learning and broad-based portfolio assessment were clearly observed." (Tanner, 2000; p. 79)

Portfolios enrich conversations about learning. Portfolios have different looks depending on purpose and audience. The audiences for classroom-based portfolios include the students themselves, their teacher(s), parents, and sometimes community members or future employers. This enhances the credibility of the process. Portfolios encourage students to show what they know and provide a supportive framework within which learning can be documented. Using portfolios in classrooms as part of the organising and collecting of evidence prepares students to present their learning and to engage in conversation (sometimes in writing, sometimes orally or through presentations) about their learning. Julius (2000) asserts that knowing they will be showing portfolios to someone whose opinion they care about engenders "accountability and a sense of responsibility for what was in the portfolios" (p. 132). Willis (2000) notes, "This formal conferencing process challenges students to be more accountable to an authentic audience outside of their classroom and generally improves the quality..." (p. 47)

When individual student portfolios are scored and graded, the power of student choice, ownership, and responsibility may be diminished. Willis (2000) states that rewards and sanctions are… "antithetical to learner centred goals of true portfolio culture" p. 39 (for a discussion of the impact of rewards see Kohn, 2000). Willis (2000) refers to student, teacher and institutional learning after examining how Kentucky's Grade 12 writing portfolios have influenced senior's writing instruction and experiences, affected students' skills and attitudes about writing, and influenced graduates' transition to college writing, He collected data using exploratory surveys with 340 seniors, interviewing 10 students who graduated and continued on with their education at the college levels, and conducted a document analysis of writing portfolios produced in senior English classes as well as samples of writing submitted in college composition courses. Willis notes the self assessments demonstrated little awareness of the standards in effect in the learning environment. Willis (2000) reports that a statistical analysis of 340 students showed that students disregarded the worth of the portfolio process to the same extent they had been disappointed with the scores received. As a result, Willis (2000) recommends that students have more experience scoring their own work. Thome (2001) studied the impact of using writing criteria on student learning and found that when students were aware of the criteria for success their writing improved. Similarly, Young (2001) found that the use of rubrics motivate, lend encouragement to learners to improve, and provide a means for giving specific feedback.

4.4 Portfolios: Feedback that Supports Learning

When portfolio are accompanied by criteria that are written in language students can understand, describe growth over time, as well as indicate what is required to achieve success they can be used by students to guide their learning with on-going feedback as they create their portfolios. There is a vast amount of research concerning the impact of feedback on student's learning. There is evidence that specific feedback is essential for learning (Black & Wiliam, 1998; Caine & Caine, 1991; 1999; Carr & Kemmis, 1986; Crooks, 1988; Dewey, 1933; Elbow, 1986; Hattie, in press; Sadler, 1989b; Senge, 1990; Shepard, 2000; Stiggins, 1996; Sylwester, 1995). Sutton (1997) and Gipps & Stobart (1993) distinguish between descriptive and evaluative feedback. Descriptive feedback serves three goals: 1) it describes strengths upon which further growth and development can be established; 2) it articulates the manner in which performance falls short of desired criteria with an eye to suggesting how that can be remediated; and 3) it gives information that enables the learner to adjust what he or she is doing in order to get better. Specific descriptive feedback that focuses on what was done successfully and points the way to improvement has a positive effect on learning (Black & Wiliam, 1998; Butler, 1987, 1988; Butler & Nisan, 1986; Butterworth & Michael, 1975; Fuchs & Fuchs, 1985; Kohn, 1993). Descriptive feedback comes from many sources. It may be specific comments about the work, information such as posted criteria that describe quality, or models and exemplars that show what quality looks like and the many ways in which it can be expressed. Evaluative feedback, particularly summary feedback, is very different. It tells the learner how she or he has performed as compared to others or to some standard. Evaluative feedback is highly reduced, often communicated using letters, numbers, checks, or other symbols. It is encoded, and is decidedly not "rich" or "thick" in the ways suggested of descriptive feedback above. This creates problems with evaluative feedback for students -- particularly for students who are struggling. Beyond the obvious disappointment of the inability of summary feedback to address students' needs or the manner in which further growth and development can be realised, there are also problems that affect students' motivation to engage in learning. Students with poor marks are more likely to see themselves as failures. Students who see themselves as failures may be less motivated and therefore less likely to succeed as learners (Black & Wiliam, 1998; Butler, 1988; Kamii, 1984; Kohn, 1993; Seagoe, 1970; Shepard & Smith, 1986a, 1987; Schunk, 1996). Involving students in assessment increases the amount of descriptive specific feedback available to learners while they are learning. Limiting specific feedback limits learning (e.g. Black & Wiliam, 1998; Hattie, in press; Jensen, 1998; Sadler, 1989b).

Joslin (2002) studied the impact of criteria and rubrics on the learning of students in fifth and sixth grade (approximately 9 – 12 years of age). He found that when students use criteria in the form of a rubric that describes development towards success, students are better able to identify strengths and areas needing improvement. Joslin (2002) found that using criteria and rubrics affect student's desire to learn in a positive way and expand their ability to assess and monitor their own learning. He notes that when scores alone were used, students who did not do well also did not know how to improve performance in the future. When students and teachers used the rubrics that described success they were able to talk about what they had done well and what they needed to work on next. Joslin (2002) states, 'Students from the treatment group who received the rubric were aware of how well they would do on an assignment before being marked. The reason for their understanding was based on comments indicating they could check out descriptors they had completed. They were also able to identify what was needed to complete the task appropriately, indicating an awareness of self-monitoring and evaluation. In the comparison group students' comments reveal a lack of understanding of how they were evaluated. Students also indicated they would try harder to improve their grade next time but were unaware of what they needed to do to improve." (p. 41). He concludes by writing, "This research study has indicated a positive relationship between the use of a rubric and students desire to learn." (p. 42). When students have clear criteria, feedback can be more descriptive and portfolios can better support learning.

4.5 Portfolios and Self-Reflection

Meta-cognitive skills are supported and practised during portfolio development as students reflect on their learning and select work samples, put work samples in the portfolio, and prepare self-assessments that explain the significance of each piece of work. Portfolio construction involves skills such as awareness of audience, awareness of personal learning needs, understanding of criteria of quality and the manner in which quality is revealed in their work and compilations of it as well as development of skills necessary to complete a task (e.g. Duffy, Jones & Thomas, 1999; Mills-Court & Amiran, 1991; Yancey, 1997).

Students use portfolios to monitor progress and to make judgements about their own learning (Julius, 2000). Julius (2000) examined elementary students' perceptions of portfolios by collecting data from 22 students and their teachers from two third grade classrooms. Data collection included student and teacher interviews, observation of student-teacher conferences, portfolio artefacts, teacher logs and consultations with teachers. Portfolios were found to contribute to student's ability to reflect upon their work and to the development of students' sense of ownership in the classroom. Julius (2000) reports, "Results of this study indicated that students used portfolios to monitor their progress, students made judgements based on physical features, choice was a factor in the portfolio process, and, instructional strategies supported higher order thinking." (p. vii) As students become more used to using the language of assessment in their classroom as they set criteria, self assess and give peers descriptive feedback, they become better able to use that feedback to explain the significance of different pieces of evidence and later to explain their learning to parents and others.

One key aspect of classroom portfolios is students' selecting evidence from multiple sources and explaining why each piece of evidence needs to be present – what it shows in terms of student learning and the manner in which it addresses the audience and the purpose of the portfolio. Portfolios communicate more effectively when the viewer knows why particular evidence has been included. Students who are involved in classroom assessment activities such as developing criteria use the language of assessment as they develop criteria and describe the levels of quality on the way. Practice using the language of assessment prepares students to reflect. Their self-assessments become more detailed and better able to explain what evidence different pieces of evidence show. Initially, work is selected for reasons as "It was the longest thing I wrote." or "It got the best grade." Over time notions of quality become more sophisticated and citing specific criteria in use in the classroom and the manner in which evidence in the portfolio address those criteria. The capacity to do so is essential to high performance learning. Bintz and Harste (1991) explain, "Personal reflection required in portfolio evaluation increases students' understanding of the processes and products of learning..."

4.6 Portfolios: Teachers as Learners

Just as students learn by being involved in the portfolio process, so do teachers. There are five key ways teachers learn through portfolio use:
1. Teachers learn about their students as individuals by looking at their learning represented in the portfolios.
2. Teachers learn about what evidence of learning can look like over time by looking at samples of student work.
3. Teachers form interpretative communities that most often have higher standards and more consistently applied standards (both from student to student and from teacher to teacher) for student work than was the case before entering into the development of portfolio systems.
4. Teachers challenge and enrich their practice by addressing the higher expectations of student learning with classroom activities that more effectively address that learning and
5. Teachers learn by keeping portfolios themselves to show evidence of their own learning over time.

Tanner (2000) says that while there was some direct student learning from portfolio assessment, perhaps the "greater learning came from post-assessment teachers who created a better climate for writing and learning" (p. 63). Teachers, who knew more about learning returned to classrooms prepared to, "impact subsequent student cohorts." (Tanner, 2000; p. 71). Tanner (2000) describes the learning - for students as well as their instructors – that emerges as students are involved in a school-wide portfolio process. Based on interviews with key informants he describes the changes that have occurred since 1986 that indicate positive changes in student attitude, choice, ownership, engagement as well as changes in teachers' understanding and knowledge, and changes throughout the college. Herman, Gearhart and Aschbacher (1996) also report that portfolio use results in learning by improving teaching. The example given is Aschbacher's (1993) action research which showed two thirds of teachers reporting substantial change in the way they thought about their teaching, two thirds reporting an increase in their expectations for students, and a majority found that alternative assessments such as portfolios reinforced the purpose or learning goals.

There is increasing attention being paid to the process of examining student work samples as part of teachers' professional learning and development (e.g. Blythe et al., 1999; Richards, 2001; MAPP, 2002). This is a natural outgrowth of:

- conversations amongst teachers (e.g. www.lasw.org; Blythe et al., 1999),
- school improvement planning processes (e.g. B.C. School Accreditation Guide, 1990; Hawai`s Standards Implementation Design Process, 2001),
- large-scale assessments (e.g. B.C. Ministry of Education, 1993; Fritz, 2001; Willis, 2000), and
- school-level work with parents to help them understand growth over time (Busick, 2001; Cameron, 1991).

There are multiple reasons teachers examine student work samples by themselves or with colleagues as part of their own professional learning:

1. Understanding individual students' growth and development to inform students, teachers, and parents about learning.
2. Increasing expectations of students (as well as the system that serves them) through encounters with student work that reveal capacities greater than previously believed.
3. Making expectations for student performance more consistent, both across teachers and across students.
4. Understanding next teaching steps by examining student work with colleagues analysing strengths, areas needing improvement and next teaching steps.
5. Learning how to evaluate work in relation to unfamiliar standards fairly by comparing samples from students within a school.
6. Gaining a better understanding of development over time by looking at samples of student work and comparing them to published developmental continuums.
7. Developing a common understanding of standards of quality by looking at samples of student work in relation to standards.
8. Learning to use rubrics from large-scale or district assessments to analyse work samples.
9. Considering student achievement over time within a school or across schools in a district.
10. Informing the public of the achievement levels of groups students.

Teachers may examine student work from individuals, groups of students, multiple classes of students, or from different grade levels in different subject areas. Blythe et al. (1999) describe different ways to involve teachers in examining student work. Parents are also beginning to be invited to join the process (e.g. BC Min of Ed. School Accreditation Process, 1990).

4.7 Portfolios: Parents and Other Community Members as Learners

Portfolios can inform different stakeholders of ongoing growth and development (e.g. Danielson, 1996; Klenowski, 2000; McLaughlin & Vogt, 1996; Shulman, 1998; Wolf, 1996; Zeichner & Liston, 1996). Efforts to include parents and others as assessors of student learning have revealed a further power of portfolios. Not only do parents come to a fuller understanding of their children's learning, they better appreciate the goals and instructional approaches of the learner and the teacher(s). This in turn makes them more effective partners in their children's learning and ensures their support for teachers' efforts at innovation and change (Howard & LeMahieu, 1995; Joslin, 2002b). Conversations with parents, teachers, and children with portfolios as an evidentiary basis provide a more complete picture of children's growth and understanding than standardised test scores. They also provide ideas so parents can better support their children's learning in and out of school, so teachers can better support the learner in school, and so the learners can support themselves as they learn. Further, portfolios and the conversations that take place in relation to them, can promote the involvement of all the members of the learning community in educating children (e.g. Gregory et al., 2001; Fu & Lamme, 2002).

4.8 Portfolios: Schools and Systems Learn

Schools learn (e.g. Costa & Kallick, 1995; Fullan, 1999; Schlechty, 1990; Schmoker, 1996; Senge, 2000; Sutton, 1997) and need assessment information in order to continue to learn and improve. Portfolios and other work sample systems can help schools both learn and show their learning. Systems learn (e.g. Senge, 1990) and need reflective feedback to help them continue to improve. They need assessment information. Portfolios can be a part of the evidence collected both to support and to substantiate the learning and are increasingly used for assessment of students, teachers, schools, districts, and educational jurisdictions such as provinces or states (Darling-Hammond, 1997; Fritz, 2001; Gillespie et al., 1996; Millman, 1997; Ryan & Miyasaka, 1995). When it comes to systems learning, numerous researchers have made recommendations based on their experiences with portfolios for district and cross-district assessments. Reckase (1995), for example, recommends a collection that represents all the tasks students are learning including both final and rough drafts. Fritz (2001) reviewed the Vermont large-scale portfolio assessment program and notes that since it began in 1989 there have been numerous studies and recommendations leading towards ongoing improvements in the design and scoring as well as the way data is used to improve the performance of schools. Portfolios are collected from all students, scored at the school level and then a sampling of portfolios is also scored at the state level. Kentucky has had a portfolio component in its assessment system since 1992 when the Kentucky Educational Reform Act became law (See Redfield & Pankratz, 1997 or Willis, 2000 for a historical overview). Like Vermont, the Kentucky mandated performance assessment has evolved over time with changes being made to the portfolio contents as well as the number of work samples required. Overtime some multiple choice test items and on-demand writing have been included (Lewis, 2001). The state of Maine has an on-going portfolio project that is developing tasks and scoring rubrics for use in classrooms as part of the Comprehensive Local Assessment System districts need to have in place by 2007. The New Standards Project co-ordinated and reported by Resnick and Resnick (1993) looked at samples of student work in Mathematics and English/Language Arts from almost 10,000 grade-4 students.

5. DISCUSSION

Barth (2000) has made the point that in 1950 students graduated from high school knowing 75% of what they needed to know to succeed. In 2000, students graduated with 2% of what they needed to know because 98% of what they needed to know to be successful was not yet known. This fact alone fundamentally changes the point of schooling. Today a quality high school education that provides these new basic skills is a minimum. Even more than this, a quality high school education must equip the learner to continuously grow, develop and learn throughout his or her lifetime. Post-secondary education can strive to do no less. For example, the Globe and Mail, a national Canadian newspaper noted, "employers' relentless drive to avoid the ill-educated. (March 1, 1999)." They went on to note that jobs for people with no high school diplomas fell 27%. In 1990 employees with post-secondary credentials held 41% of all jobs while in 1999 that had risen to 52% of all jobs. The trend is expected to continue and the gap widen. Government commissions, business surveys, newspaper headlines and the help wanted advertisements all testify to the current reality – wanted: lifelong learners who have new skills basic to this knowledge age – readers writers, thinkers, technologically literate, and able to work with others collaboratively to achieve success. We can't prepare students to be lifelong learners without changing classroom assessment. Broadfoot (1998) puts it this way, "the persistence of approaches to assessment which were conceived and implemented in response to the social and educational needs of a very different era, effectively prevents any real progress." (p. 453) Traditional forms of assessment were not conceived without logic, they were conceived to respond to an old, now outdated, logic.

Meaningful assessment reform will occur when
- students are deeply involved in the assessment process;
- evidence of learning is defined broadly enough for all learners to show what they know;
- classroom assessment is understood to be different than other kinds of assessment;
- an adequate investment in assessment for learning is made; and,
- a proper balance is achieved between types of assessment

Accountability for individual student learning involves looking at the evidence with learners, making sense of it in terms of student strengths, areas needing improvement, and helping students learn ways to self-monitor their way to success. Classroom assessment will achieve its primary purpose of supporting student learning when it is successfully used to help students learn more and learn ways to lead themselves to higher levels of learning. Portfolio assessment can play a significant role.

In our experience, two things have invariably been realised through the "assessment conversations" that are entered into by teachers in the development of portfolio systems. Both of these outcomes greatly enhance the intellectual and human capital of the systems and contribute to the potential for their improved performance. First, all who participate in the development of portfolio systems leave with higher and more clearly articulated aspirations for student performance. This should not be surprising, as the derivation of criteria and expectations for quality in performances is essentially additive. One professional sees certain things in the student work while the next recognises some of these (but perhaps not all) and adds some more. These assessment conversations proceed until the final set of aspirations (criteria of quality) is far greater than the initial one or that of any one member of the system at the outset. The second effect of these assessment conversations is that a shared interpretative framework for regarding student work emerges. The aspirations and expectations become commonly understood across professionals and more consistently applied across students. Again, the nature of these conversations (long term shared encounters and reflections) intuitively supports this outcome.

These two outcomes of assessment conversations -- elevated aspirations and more consistently held and applied aspirations -- are key ingredients in a recipe for beneficial change. Educational research is nowhere more compelling than in its documentation of the relationship between expectations and student accomplishment. Where expectations are high and represent attainable yet demanding goals, students strive to respond and ultimately achieve them. These assessment conversations, focused upon student work produced in response to meaningful tasks provide powerful evidence that warrants the investment in the human side of the educational system.

It is for these reasons that we are optimistic about the place of portfolios in reform in North America. Yet, that said, portfolios are not mechanical agents of change. We do not accept the logic that says that the testing (however new or enlightened) coupled with North America version of accountability will motivate increased performance. In fact, we find it a cynical argument presuming as it does that all professionals in the system could perform better but for reasons (that will be eliminated by the proper application of rewards and sanctions) they have simply chosen not to. However, our experience also suggests that in order for the full potential of assessment development or teacher and student engagement in rich and rewarding assessment tasks to be realised, it must be approached in a manner consistent with the understandings developed here.

Portfolios pose unique challenges in large-scale assessment. Involving students in every part of the portfolio process is critical to its success as a learning and assessment tool. Choice and ownership, thinking about their thinking, and preparing evidence for an audience whose opinion they care about are key aspects of portfolio use in classrooms. These critical features risk being lost when the portfolio contents and selection procedures are dictated from outside the classroom for accountability purposes. Without choice and student ownership, portfolios may be limited in their ability to demonstrate student learning. This may mean that large-scale portfolio assessment may become a barrier to individual student learning. However, using portfolios for large-scale assessment (when done well) can potentially support system learning in at least these ways:

- Facilitating a better understanding of learning and achievement trends and patterns over time
- Informing educators about learning and assessment as they analyse resulting student work samples
- Enhancing professionals' expectations for students (and themselves as facilitators of student learning) as a result of working with learner's portfolios

- Making it possible to assess valued outcomes that are well beyond the reach of other means of assessment
- Informing educators' understandings of what learning looks like over time as they review collections of student work samples
- Helping students to understand quality as they examine collections of student samples to better understand the learning and what quality can look like
- Assisting educators and others to identify effective learning strategies and programs

Purpose is key. Whose learning is intended to be supported? Student? Teacher? School? System?

Assessments without a clear purpose risk muddled methods, procedures, data, and findings (e.g. Chan, 2000; Paulson, Paulson, & Meyer, 1991; Stiggins, 2001). For example, one group indicated that the jurisdiction could use portfolios to assess individual student achievement, teaching, educators, schools, and provide state level achievement information (see for example, Richard, 2001). This is true but different portfolios would be required or the purpose could be confused, the methods inappropriately used, the procedures incorrect, the resulting portfolios likely inappropriate to the stated purposes and the findings inaccurate. When the purpose and audience shifts, the portfolio design, content, procedures, and feedback need to be realigned. If the purpose for collecting evidence of learning and using portfolios is to support student learning then it may not be necessary for portfolios to be evaluated, scored or graded. If the purpose for collecting evidence of learning and using portfolios is to support educators (and others) as they learn and seek to improve system performance then portfolios will be standardised to some necessary degree, evaluated and scoring results made available to educators and others.

REFERENCES

Anson, C. M., & Brown, R. L. (1991). Large scale portfolio assessment: Ideological sensitivity and institutional change. In P. Belanoff & M. Dickson (Eds), *Portfolios : Process and Product* (pp. 248-269). Portsmouth, NH: Boynton/Cook

Anthony, R., Johnson, T., Mickelson, N., & Preece, A. (1991). *Evaluating Literacy: A Perspective for Change*. Portsmouth, NH: Heinemann.

Arter, J. A., & Spandel, V. (1992). Using portfolios of student work in instruction and assessment. *Educational Measurement: Issues and practice, 11* (1), 36-44.

Arter, J., Spandel, V., & Culham, R. (1995). *Portfolios for assessment and instruction*. Greensboro, NC: ERIC Clearinghouse on Counselling and Student Services (ERIC Document Reproduction Service No. ED 388 890).

Aschbacher, P. R. (1993*).* *Issues in innovative assessment for classroom practice: Barriers and facilitators.* (CSE Tech Rep No. 359). Los Angeles, CA: University of California, National Center for Research on Evaluation, Standards and Student Testing.

Baron, J.B. (1983). *Personal communication to the authors.*

Barth, R. (2000). *Learning by Heart.* San Francisco, CA: Jossey-Bass Publishers.

Benoit, J., & Yang, H. (1996). A redefinition of portfolio assessment based upon purpose: Findings and implications from a large-scale program. *Journal of Research and Development in Education, 29* (3), 181-191.

Berryman; L., & Russell, D. (2001). Portfolios across the curriculum: Whole school assessment in Kentucky; *English Journal, Urbana, 90* (6), 76. High school edition.

Biggs, J. (1995). Assessing for learning: some dimensions underlying new approaches to educational assessment, *Alberta Journal of Educational Research, 41*, 1-17.

Bintz, W., & Harste, J. (1991). Whole language assessment and evaluation: The future. In B. Harp (Ed.), *Assessment and evaluation in whole language programs* (pp. 219-242). Norwood, MA: Christopher Gordon.

Black, P., & Wiliam, D. (1998). Assessment and classroom learning. *Assessment in Education. 5* (1), 7-75.

Blythe, T., et al. (1999). *Looking Together at Student Work.* New York, NY: Teachers College Press.

B.C. Ministry of Education. (1990, 2000). *Primary Program.* Victoria, B.C.: Queens' Printer.

B.C. Ministry of Education. (1990). *School Accreditation Guide.* Victoria, B.C.: Queens' Printer.

B.C. Ministry of Education. (1993). *British Columbia Provincial Assessment of Communication Skills: How well do British Columbia Students Read, Write, and Communicate?* Victoria, B.C.: Queens' Printer.

Broad, R. L. (1994). "Portfolio Scoring": A contradiction in terms. In L. Black, D. A. Daiker, J. Sommers, & G. Stygall (Eds), *New Directions in Portfolio Assessment: Reflective practices, critical theory, and large scale assessments* (pp. 263-276). Portsmouth, NH: Boynton/Cook.

Broadfoot, P. (1998). Records of achievement and the learning society: a tale of two discourses. *Assessment in Education: Principles, Policy and Practice, 5* (3), 447-477.

Busick, K. (2001). *In Conversation.* Kaneohe, Hawaii.

Butler, R. (1987). Task-involving and ego-involving properties of evaluation: Effects of different feedback conditions on motivational perceptions, interest and performance. *Journal of Educational Psychology 79* (4), 474-482.

Butler, R. (1988). Enhancing and undermining intrinsic motivation: The effects of task-involving and ego-involving evaluation on interest and performance. *Journal of Educational Psychology 58,* 1-14.

Butler, R., & Nisan, M. (1986). Effects of no feedback, task-related comments and grades on intrinsic motivation and performance. *Journal of Educational Psychology 78* (3), 210-216.

Butterworth, R. W., & Michael, W. B. (1975). The relationship of reading achievement, school attitude, and self-responsibility behaviors of sixth grade pupils to comparative and individuated reporting systems: implication of improvement of validity of the evaluation and pupil performance. *Educational and Psychological Measurement (35),* 987-991.

Caine, R. N., & Caine, G. (1991). *Making Connections: Teaching and the Human Brain.* Alexandria, Virginia: ASCD.

Calfee, R. C. & Freedman, S. W. (1997). Classroom writing portfolios: Old, new, borrowed, blue. In R. Calfee & P. Perfumo (Eds), *Writing portfolios in the classroom: Policy and practice, process and peril.* (pp. 3-26). Hillsdale, NJ: Erlbaum.

Callahan, S. F. (1995). *State mandated writing portfolios and accountability: an ethnographic case study of one high school English department.* Unpublished doctoral dissertation. University of Louisville, Kentucky.
Cameron, C. (1991). *In conversation.* Sooke, B.C.
Cameron, C., Tate, B., Macnaughton, D., & Politano, C. (1998). *Recognition without Rewards.* Winnipeg, Man.: Peguis Publishers.
Carney, J. M. (2001) *Electronic and traditional paper portfolios as tools for teacher knowledge representation.* Doctoral Dissertation. University of Washington.
Carr, W., & Kemmis, S. (1986). *Becoming Critical: Education, knowledge, and action research.* London: The Falmer Press.
Chan, Yat Hung (2000). *The assessment of self-reflection in special education students through the use of portfolios.* Doctoral dissertation University of California, Berkeley.
Condon, W, & Hamp-Lyons, L. (1994). Maintaining a portfolio-based writing assessment: Research that informs program development. In L. Black, D. A. Daiker, J. Sommers, & G. Stygall (Eds.), *New Directions in Portfolio Assessment: Reflective practices, critical theory, and large scale assessment* (pp. 277-85). Portsmouth, NH: Boynton/Cook.
Costa, A., & Kallick, B. (1995). *Assessment in the Learning Organization.* Alexandria, Va.: ASCD.
Costa, A., & Kallick, B. (Eds.). (2000). *Assessing and Reporting on Habits of Mind.* Alexandria, VA: ASCD.
Covington, M. (1998). *The Will to Learn: A Guide for motivating Young People.* Cambridge, UK: Cambridge University Press.
Crooks, T. (1988). The impact of classroom evaluation on students. *Review of Educational Research, 58* (4), 438-481.
Danielson, C. (1996). *Enhancing professional practice: A framework for teaching.* Virginia: ASCD.
Darling-Hammond, L. (1997). *The Right to Learn.* San Fransisco, CA: Jossey-Bass Publishers.
Darling-Hammond, L. (1997). Toward What End? The Evaluation of student learning for the improvement of teaching. In J. Millman (Eds.), *Grading Teachers, Grading School: Is Student Achievement a Valid Evaluation Measure?* Thousand Oaks, CA: Corwin Press, Inc.
Davies, A. (2000). *Making Classroom Assessment Work.* Merville, B.C.: Connections Publishing.
Davies, A., Cameron, C., Politano, C., & Gregory, K. (1992). *Together Is Better: Collaborative Assessment, Evaluation, and Reporting.* Winnipeg, Man.: Peguis Publishers.
Deci, E. L., & Ryan, R. M.. (1985). *Intrinsic motivation and self-determination in human behavior.* New York. Plenum Press.
DeCharms, R. (1968). *Personal Causation: The internal affective determinants of behaviour.* NY: Academic Press.
DeCharms, R. (1972). Personal causation training in schools. *Journal of Applied Social Psychology. 2,* 95-113.
DeVoge, J. G. (2000). *The measurement quality of integrated assessment portfolios and their effects on teacher learning and classroom instruction.* Duquesne University, Doctoral dissertation Dissertation Abstracts.
Dewey, J. (1933). *How We Think: A Restatement of the Relation of Reflective Thinking To the Educative Process.* Lexington, Mass.: Heath.
Duffy, M. L., Jones, J. & Thomas, S. W. (1999). Using portfolios to foster independent thinking. *Intervention in School and Clinic, 35* (1).

Durth, K. A. (2000). *Implementing portfolio assessment in the music performance classroom.* Doctoral dissertation Columbia University Teachers College.

Earl, L. M., & LeMahieu, P. G. (1997a). Rethinking assessment and accountability. In A. Hargreaves (Ed.), *Rethinking educational change with heart and mind.* (1997). Yearbook of the Association for Supervision and Curriculum Development. Association for Supervision and Curriculum Development. Alexandria, VA.

Elbow, P. (1986). *Embracing Contraries: Explorations in Learning and Teaching* (pp. 231-232). New York: Oxford University Press.

Elbow, P. & Belanoff, P. (1986). Portfolios as a Substitute for Proficiency Examinations. *College Composition and Communication, (37),* 336-339.

Elbow, P., & Belanoff, P. (1991). State University of New York at Stony Brook. Portfolio-based Evaluation Program. In P. Belanoff & M. Dickson (Eds.), *Portfolios: Process and Product* (pp. 3-16). Portsmouth, NH: Boynton/Cook.

Flood, J., & Lapp, D. (1989). Reporting reading progress: A comparison portfolio for parents. *The Reading Teacher,* 508-514.

Fredericksen, J. R., & Collins, A. (1989). A Systems Approach to Educational Testing. *Educational Researcher. 18,* 27-52.

Fritz, C.A. (2001). *The level of teacher involvement in the Vermont mathematics portfolio assessment.* University of New Hampshire. Unpublished doctoral dissertation.

Fu, D., & Lamme, L. (2002). Assessment through Conversation. *Language Arts, 79* (3), 241-250.

Fuchs, L. S. and Fuchs, D. A. (1985). *Quantitative synthesis of effects of formative evaluation of achievement.* Paper presented at the annual meeting of the American research association. Chicago Ill. ERIC Doc. #ED256781.

Fullan, M. (1999). *Change Forces: the Sequel.* Philadelphia, PA: George H. Buchanan.

Gardner, H. (1984). *Frames of Mind: The theory of multiple intelligences.* New York: Basic Books.

Gardner, H., & Boix-Mansilla, V. (1994). Teaching for Understanding - Within and Across the Disciplines. *Educational Leadership, 51* (5), 14-18.

Gearhart, M., Herman, J., Baker, E. L., & Whittaker, A. K. (1993). *Whose work is it?* (CSE Tech Rep No. 363). Los Angeles, CA: University of California Center for Research on Evaluation, Standards and Student Testing.

Gearhart, M., Herman, J. & Novak, J. (1994). *Toward the instructional utility of large-scale writing assessment: Validation of a new narrative rubric.* (CSE Tech Report No. 389). Los Angeles, CA: University of California, National Center for Research on Evaluation, Standards and Student Testing.

Gearhart M., & Wolfe, S. (1995). *Teachers' and Students' roles in large-scale portfolio assessment: Providing evidence of competency with the purpose and processes of writing* (p. 69). Los Angeles: UCLA/The National Centre for Research on Evaluation, Standards, and Student Testing. (CRESST).

Gillespie, C., Ford, K, Gillespie, R., & Leavell, A. (1996). Portfolio Assessment: Some questions, some answers, some recommendations. *Journal of Adolescent & Adult Literacy, 39,* 480-91.

Gipps, C. (1994). *Beyond Testing.* Washington, DC, Falmer Press.

Gipps, C., & Stobart, G. (1993). *Assessment: A Teachers Guide to the Issues* (2nd Edition). London, UK: Hodder & Stoughton.

Goleman, D. (1995). *Emotional Intelligence.* New York: Bantam Books.

Gregory, K., Cameron, C., & Davies, A. (2001). *Knowing What Counts: Conferencing and Reporting.* Merville, B.C.: Connections Publishing.

Hansen, J. (1992). Literacy Portfolios. Helping Students Know Themselves. *Educational Leadership, 49* (8), 66-68.

Hattie, J. (in press). *The Power of Feedback for Enhancing Learning.* University of Auckland, NZ.

Herman, J. L. (1996). Technical quality matters. In R. Blum & J.A. Arter (Eds.), *Performance assessment in an era of restructuring* (pp. 1-7:1 – 1-7:6). Alexandria, VA: Association for Supervision and Curriculum Development (ASCD).

Herman, J. L, Aschbacher, P. R., & Winters, L. (1992) *A Practical Guide to Alternative Assessment.* (Alexandria, VA, Association for Supervision and Curriculum Development).

Herman, J., Gearhart, M., & Aschbacher, P. (1996). Portfolios for classroom assessment: design and implementation issues. In R. Calfee & P. Perfumo (Eds.), *Writing portfolios in the classroom: policy and practice, promise and peril.* Mahwah, NY: Lawrence Erlbaum Associates, Inc.

Hoover, H. D., & Bray, G. B. (1995). *The research and development phase: Can performance assessment be cost effective?* Paper presented at the annual meeting of the American Educational Research Association. San Francisco, CA.

Howard, K., & LeMahieu, P. (1995). Parents as Assessors of Student Writing: Enlarging the community of learners. *Teaching and Change, 2* (4), 392-414.

Jensen, E. (1998). *Teaching with the Brain in Mind.* Alexandria, Virginia: ASCD.

Joslin, G. (2002). *Investigating the influence of rubric assessment practices on the student's desire to learn.* Unpublished manuscript San Diego State University.

Joslin, G. (2002b). *In conversation.* October AAC Conference. Edmonton, Alberta.

Julius, T. M. (2000). *Third grade students' perceptions of portfolios.* University of Massachusetts Amherst. Unpublished doctoral dissertation.

Kamii, C. (1984). Autonomy: The aim of education envisioned by Piaget. *Phi Delta Kappan, 65* (6), 410-415.

Kinchin, G. D. (2001). Using team portfolios in a sport education season. *Journal of Physical Education, Recreation & Dance, 72* (2), 41-44.

Klenowski, V. (2000). Portfolios: Promoting teaching. *Assessment in Education, 7* (2), 215-236.

Konet, R. J. (2001). Striving for a personal best. *Principal Leadership, 1* (6), 18-23.

Kohn, A. (1993). *Punished by Rewards: The trouble with gold stars, incentive plans, A's, praise, and other bribes.* Boston, Mass.: Houghton Mifflin Company.

Kohn, A. (2000). *The Schools our children deserve. Moving Beyond Traditional Classrooms and "Tougher Standards".* Boston, MA: Houghton Mifflin.

Koretz D., Stecher, B., Klein, S., McCaffrey, D. & Deibert, E. (1993). *Can portfolios assess student performance and influence instruction? The 1991-92 Vermont experience.* (CSE Tech Report No. 371) Los Angeles, CA: University of California, National Center for Research on Evaluation, Standards and Student Testing.

Kuhs, T. (1994). Portfolio Assessment: Making it Work for the First Time. *The Mathematics Teacher, 87* (5), 332-335.

Law, B., & Eckes, M. (1995). *Assessment and ESL.* Winnipeg, MB: Peguis Publishers.

Lazear, D. (1994). *Multiple Intelligence Approaches to Assessment: Solving the Assessment Conundrum.* Tucson, Arizona: Zephyr Press.

Le Doux, J. (1996). *The Emotional Brain.* New York: Simon and Schuster.

LeMahieu, P. G., Eresh, J. T., & Wallace, Jr., R. C. (1992b). Using Student Portfolios for public accounting. *The School Administrator: Journal of the American Association of School Administrators, 49* (11). Alexandria, VA.

LeMahieu, P. G., Gitomer, D. A., & Eresh, J. T. (1995a). Portfolios in large-scale assessment: Difficult but not impossible. Educational Measurement: Issues and Practice. *Journal of the National Council on Measurement in Education, 13* (3).
LeMahieu, P. G., Gitomer, D. A., & Eresh, J. T. (1995b). *Beyond the classroom: Portfolio quality and qualities.* Educational Testing Service. Rosedale, NJ.
LeMahieu, P. G. (1995c). *Transforming public accountability in education.* Fifth Annual Boisi Lecture. Boston College. Chestnut Hill, MA.
LeMahieu, P. G., & Eresh, J. T. (1996a). Comprehensiveness, coherence and capacity in school district assessment systems. In D. P. Wolf and J. B. Baron (Eds.), *Performance based student assessment: Challenges and possibilities.* The 95th Yearbook of the National Society for the Study of Education. Chicago, IL.
LeMahieu, P. G. (1996b). From authentic assessment to authentic accountability. In J. Armstrong (Ed.), *Roadmap for change: A briefing for the Second Education Summit.* Education Commission of the States. Denver, CO.
LeMahieu, P. G. (1996c). *From authentic assessment to authentic accountability in education.* Invited Keynote Address to the Michigan School Assessment Conference. Ann Arbor, MI. March, 1996.
Lepper, M. R., & Greene, D. (1975). Turning play into work: effects of adult surveillance and extrinsic rewards on children's intrinsic motivation. *Journal of Personality and Social Psychology, 31,* 479-486.
Lewin, K. (1952). Group decision and social change. In G. E. Swanson, T. M. Newcomb, & F. E. Hartley (Eds.), *Readings in Social Psychology.* New York: Holt.
Lewis, S. (2001). Ten years of puzzling about audience awareness. *The Clearing House, 74* (4), 191-197.
Lincoln, Y., & Guba, E. (1984). *Naturalistic Inquiry.* Beverly Hills, Calif.: Sage Publications.
MacDonald, C. (1982). "A Better Way of Reporting" B.C. *Teacher, 61,* 142-144.
Maehr, M. (1974). *Sociocultural origins of achievement.* Monterey, Calif.: Brooks/Cole.
Mager, R. F., & McCann, J. (1963). *Learner Controlled Instruction.* Palo Alto, CA: Varian Press.
Mahoney, M. J. (1974). *Cognition and Behaviour Modification.* Cambridge, MA: Ballinger.
Maine Assessment Portfolio (MAPP). (2002).
Manning, M. (2000). Writing Portfolios. *Teaching Pre-K-8, 30* (6), 97-98.
McLaughlin, M., & Vogt, M. (1996). *Portfolios in teacher education.* Newark, DE: International Reading Assocation.
Millman, J. (1997). *Grading teachers, grading schools: Is student achievement a valid evaluation measure?* CA: Corwin Press.
Mills-Court, K., & Amiran, M. R. (1991). Metacognition and the use of portfolios. In P. Belanoff & M. Dickson (Eds.), *Portfolios: Process and Product* (pp. 101-112). Portsmouth, NH: Boynton/Cook.
Novak, J. R., Herman, J. L., & Gearhart, M. (1996). Establishing Validity for Performance-Based Assessment: An Illustration for Collections of Student Writing. *Journal of Educational Research, 89* (4), 220-233.
Paulson F. L., Paulson, P. R., & Meyer, C. (1991). What makes a portfolio a portfolio? *Educational Leadership, 40* (5), 60-63.
Politano, C., Cameron, C., Tate, B., & MacNaughton, D. (1997). *Recognition without Rewards.* Winnipeg, MB: Peguis Publishers.
Potter, E. F. (1999). What should I put in my portfolio? Supporting young children's goals and evaluations. *Childhood Education, 75* (4), 210-214.
Purkey, W. & Novak, J. (1984). *Inviting School Success.* Belmont, CA: Wadsworth.

Quesada, A. (2000). Digital entrepreneurs. *Technology & Learning, 21* (1), 46.
Reckase, M. (1995). Practical Experiences in Implementing a National Portfolio Model at the High School Level. *The National Association of Secondary School Principals* (NASSP) Bulletin.
Redfield, D. & Pankratz, R. (1997). Historical Background: The Kentucky School Accountability Index. In J. Millman (Eds), *Grading Teachers, Grading Schools*. Thousand Oaks, Calif.: Corwin Press Inc.
Reese, B. F. (1999). Phenomenal Portfolios. *The Science Teacher* (pp. 25-28).
Resnick, L., & Resnick, D. (1993). National Center for Research on Evaluation, Standards, and Student testing, Project 2.3: Complex Performance Assessments: Expanding the Scope and Approaches to Assessment, Report on Performance Standards in Mathematics and English: Results from the New Standards Project Big Sky Scoring Conference. *U.S Department of Education, Center for the Study of Evaluation*, CRESST/LRDC. U. of Pittsburgh, UCLA.
Richard, A. (2001). Rural school trying out portfolio assessment. *Education Week, 2* (9), 5.
Richards, M. (2001). *In conversation*. Falmouth, ME. www.learningeffects.com
Richter, S. E. (1997). Using portfolios as an additional means of assessing written language in a special education classroom. *Teaching and Change, 5* (1), 58-70.
Ryan, & Miyasaka. (1995). Current practices in Testing and Assessment: What is driving the changes? *National Association of Secondary School Principals (NASSP) Bulletin, 79* (573), 1-10.
Sadler, R. (1989a). Specifying and promulgating achievement standards. *Oxford Review of Education, 13*, 191-209.
Sadler, R. (1989b). Formative assessment and the design of instructional systems. *Instructional Science, 18*, 119-144.
Schlechty, P. (1990). *Schools for the 21st Century* (p. 142). San Francisco: Jossey-Bass.
Schmoker, M. (1996). *Results: The key to continuous improvement*. Virginia: Association for Supervision and Curriculum Development.
Schon, D. A. (1983). *The Reflective Practitioner*. New York: Basic Books.
Schonberger, L. C. (2000). *The intentions and reported practices of portfolio use among beginning teachers*. Duquesne University. Unpublished dissertation.
Schunk, D. H. (1996). Theory and Reasearch on Student Perceptions in the Classroom. In D. H.. Schunk, & J.L. Meece (Eds.), *Student Perceptions in the Classroom* (pp. 3-23). Hillsdale, NJ: Erlbaum.
Seagoe, M. V. (1970). *The learning process and school practice*. Scranton, Penn.: Chandler Publishing Company.
Senge, P. M. (1990). *The fifth Discipline: the Art & Practice of The Learning Organization*. New York, NY: Doubleday.
Senge, P. (2000). *Schools that Learn*. NY: Doubleday.
Shepard, L. (2000). The role of assessment in a learning culture. *Educational Researcher, 29* (7), 4-14.
Shepard, L., & Smith, M. (1986a). Synthesis of Research on School Readiness and Kindergarten Retentions. *Educational Leadership. 44*, 78-86.
Shepard, L., & Smith, M. (1987). What doesn't work: Explaining policies of retention in the early grades.. *Kappan 69*, 129-134.
Shulman, L. (1998). Teacher portfolios: a theoretical activity. In N. Lyons (Ed.), *With Portfolio in Hand: Validating the new teacher professionalism* (pp 23-37). New York NY: Teachers College Press.

Sizer, T. R. (1996). *Horace's Hope: What Works for the American High School.* Boston: Houghton Mifflin Co.

Smith, A. (2000). Reflective Portfolios: Preschool Possibilities. *Childhood Education, 76* (4), 204-208.

Stiggins, R. (1996). *Student Centered Classroom Assessment.* Columbus, Ohio: Merrill Publishing.

Stiggins, R. (2001). *Student Involved Classroom Assessment.* Columbus, Ohio: Merrill Publishing.

Stiggins, R., & Davies, A. (1996). *Student involved Conferences (video).* Assessment Training Institute. Portland, Oregon.

Sutton, R. (1997). *The Learning School.* England: Sutton Publications.

Sylwester, R. (1995). *A celebration of neurons: An educators guide to the brain.* Alexandria, Virginia: ASCD.

Tanner, P. A. (2000). *Embedded assessment and writing: Potentials of portfolio-based testing as a response to mandated assessment in higher education.* Bowling Green State Universtiy. Unpublished dissertation.

Thorne, C. C. (2001). *The Effects of Classroom-Based Assessment Using an Analytical Writing Rubric on high school students' writing achievement.* Cardinal Stritch University. Unpublished Dissertation.

Tierney, R. J., Carter, M. A., & Desai, L. E. (1991). *Portfolio Assessment in the Reading-Writing Classroom.* Norwood, MA: Christopher-Gordon Publishers.

Tjosvold, D. (1977). Alternate organisations for schools and classrooms. In D. Bartel, & L. Saxe (Eds.), *Social psychology of education: Research and theory.* New York: Hemisphere Press.

Tjosvold D., & Santamaria, P. (1977). *The effects of cooperation and teacher support on student attitudes toward classroom decision-making.* Paper presented at the meeting of the American Educational Research Association, March 1977, New York.

Valdez, P. S. (2001). Alternative Assessment: A monthly portfolio project improves student performance. *The Science Teacher.*

Walters, J., Seidel, S., & Gardner, H. (1994). Children as Reflective Practitioners. In K.C.C. Block, & J. N. Magnieri (Eds.), *Creating Powerful Thinking in Teachers and Students.* New York: Harcourt Brace.

Weaver, S. D. (1998). *Using portfolios to assess learning in chemistry: One schools story of evolving assessment practice.* Unpublished Doctoral Dissertation: Virginia Polytechnic Institute and State University.

White, E. M. (1994a). *Teaching and Assessing Writing.* San Fransisco, CA: Sage.

White, E. M. (1994b). Portfolios as an assessment concept. In L. Black, D. A. Daiker, J. Sommers, & G. Stygall (Eds.), *New directions in portfolio assessment: Reflective practices, critical theory, and large scale assessment* (pp. 25-39). Portsmouth, NH: Boynton/Cook.

White, E. M. (1995). *Teaching and assessing writing.* San Fransisco, CA: Sage.

Wiggins, G. (1993). *Assessing Student Performance: Exploring the Purpose and Limits of Testing.* San Francisco, Calif.: Jossey-Bass Publishers.

Willis, D. J. (2000). *Students perceptions of their experiences with Kentucky's mandated writing portfolio.* Unpublished doctoral dissertation. University of Louisville.

Wolf, D., Bixby, J., Glenn, J. & Gardner, H. (1991). To use their minds well: Investigating new forms of student assessments. In G. Grant (Ed.), *Review of Research in Education* (Vol 17, pp. 31-74). Washington, DC: American Educational Research Association.

Wolf, K. (1996). Developing an effective teaching portfolio. *Educational Leadership, 53* (6), 34–37.
Wong-Kam, J., Kimura, A., Sumida, A., Ahuna-Ka`ai, J., & Hayes Maeshiro, M. (2001). *Elevating Expectations*. Portsmouth, NH: Heinemann.
www.state.me.us/education/salt.localassess.htm : State of Maine's ongoing portfolio project 's website.
Yancey, K. B. and Weiser, I. (1997). *Situating Portfolios: Four perspectives*. Logan, UT: Utah State University Press.
Young, C. A. (2001). *Technology integration in inquiry-based learning: An evaluation study of a web-based electronic portfolios*. University of Virginia.
Zeichner, K. M. & Liston, D. P. (1996). *Reflective Teaching: an Introduction*. New York, Lawrence Erlbaum Associates.

Students' Perceptions about New Modes of Assessment in Higher Education: a Review

Katrien Struyven, Filip Dochy & Steven Janssens
University of Leuven, Department of Instructional Science, Centre for Research on Teacher and Higher Education, Belgium

1. INTRODUCTION

Educational innovations in instruction and assessment have been overwhelming during the latest decade: new teaching methods and strategies are introduced in teacher and higher education and teaching practice, the latest technologies and media are used, and new types and procedures of assessment are developed and implemented. Most of these innovations are inspired by constructivist learning theories, in which the learner is an active partner in the process of learning, teaching and assessment. This belief in the active role of the student in instruction and assessment and the finding of Entwistle (1991) that it are students' perceptions of the learning environment that influence how a student learns, not necessarily the context in itself, both gave rise to this review study. Reality per se is often not sufficient to fully understand student learning and accompanying assessment processes. 'Reality as experienced by the student" has in this respect an important additional value. It is this second-order perspective (Van Rossum & Schenk, 1984), that is the primary concern of this review on new modes of assessment. Our purpose is to overview the research and literature on students' perceptions about assessment, with the aim to achieve a better understanding of students' perceptions about assessment in higher education and to gain insight into the potential impact of these perceptions on student learning, and more broadly, the learning- teaching environment. Following questions were of special interest to this review: (1) what are the influences of the (perceived) characteristics of assessment on students' approaches to learning, and vice versa, (2) what are students' perceptions about different novel assessment formats and methods, and (3) what are the influences of these students' perceptions about assessment on student learning?

2. METHODOLOGY FOR THE REVIEW

The Educational Resources Information Center (ERIC), the Web of Science, PsychINFO and Current Content, were searched online for the years 1980 until now. The keywords "student* perception*" and "assessment" were combined. This search yielded 508 hits in the databases of ERIC and PsycINFO and 37 hits within the Web of Science. When this search was limited with the additional keyword "higher education", only 171 hits and 10 hits respectively remained. Relevant documents were selected and searched for in the libraries and the e- library of the K.U. Leuven. For the purpose of this review on students' perceptions about assessment in higher education, 35 documents met our criteria. Within these selections of literature, 36 empirical studies are discussed. For a summary of this literature, we refer to the overview, presented in table 1. Theoretical and empirical articles are both included. Using other literature reviews as a guide (Topping, 1998; Dochy, Segers, Gijbels & Van den Bossche, 2002), we defined the characteristics central to this review and analysed the empirical articles according to these characteristics. First, a specific code is given to each article, for example: 1996/03/EA. This code refers to the publication year/ number/ publication type (EA: empirical article/ TA: theoretical article/ CB: chapter of book). Second, the author(s) and title of the publication are presented. Further, the overview reports on the following characteristics of the reviewed research: (1) the content of the study, (2) the type and method of the investigation, (3) the subjects and the type of education in which the study is conducted, (4) the number of subjects in the experimental and control group, (5) the most important results that were found, (6) the independent and (8) dependent variables studied, (7) the treatment which was used, and (9) the type and (10) method of analyses reported in the research. Both qualitative and quantitative investigations are discussed. Because of the large diversity of research that was found on this particular topic (e.g. exploratory studies, experiments, surveys, case studies longitudinal studies, cross- section investigations, qualitative interpretative research and quantitative research methods), a narrative review is conducted here. In a narrative review, the author tries to make sense of the literature in a systematic, creative and descriptive way (Van IJsendoorn, 1997). To prevent bias, because of the more intuitive nature of the narrative review, we reported on the procedure and criteria used to locate the studies and we described the methodological issues of the research as completely and objectively as possible.

Code	Author	Title	Content of the study	Research-type and -method	Subjects (N total) and type of education	N exp and N control
2001/01/EA	I. Treadwell & S. Gobler	Students' perceptions on skills training in simulation.	Students' experiences with acquiring practical skills in a skills laboratory and the impact that acquiring these skills has on their clinical practice were explored and described.	Qualitative, investigative, descriptive and contextual design. Focus group discussions.	196 fourth year undergraduate medical students in a problem-oriented medical curriculum.	/
2001/02/EA	Gary J. Mires, Miriam Friedman Ben-David, Paul E. Preece & Brenda Smith	Educational benefits of student self-marking of short-answer questions.	Pilot study to evaluate feasibility and reliability of undergraduate medical student self-marking of degree written examinations, and to survey students' opinions regarding the process.	Paper with 4 constructed response questions was administered to volunteers under examination conditions., afterwards they attended a self-marking session. Evaluation form for opinions.	119 second year students volunteered for the examination, 99 of them attended the self-marking session.	/
2001/03/TA	M. Friedman B-D, M. H. Davis, R. M. Harden, P. W. Howie, J. Ker, M. J. Pippard	AMEE: Medical Education Guide No. 24: Portfolios as a method of student assessment.	A guide to inform medical teachers about the use of portfolios for student assessment. It provides a background to the topic, the Dundee case study and a practical guide for those wishing to design and implement portfolio assessment in their own institution. Reasons for using portfolios for assessment purposes include the impact that they have in driving student learning and their ability to measure outcomes such as professionalism that are difficult to assess using traditional methods.			
2001/04/EA	Maggie Challis	Portfolios and assessment : meeting the challenge.	Editorial on portfolios assessment. A small literature review is given.			
2001/05/EA	Mien Segers & Filip Dochy	New assessment forms in problem-based learning : the value- added of the students' perspective.	Quantit. and qualitative data gathered from a research project which is focused on different quality aspects of two new assessment forms in problem-based learning.	S1: written examination (overall test) and students' perc's on this assessment mode. S2: peer assessment and idem.	S1: TOC1: 34, and TOC2: 45, st. evaluation questionnaire: 100, interviews: 33 students. S2: 27 students in two groups.	/

Outcomes/ results/ conclusions	Independent variable	Treatment	Dependent variable	Type of analysis	Method of analysis/ Value of statistics reported
Attitudes, knowledge and skills are interrelated and contribute to an enhanced process of learning. The interrelated learning process has a positive effect when students progress from the skills laboratory to clinical practice. Students made valuable recommendations that should contribute to the optimal use of the skills laboratory.	/	Discussions (two questions) in three focus groups (n= 24 students), field notes were taken and audiotape recordings. Successive discussions were conducted until repetition of themes (= saturation of data).	Students experiences in a skill laboratory.	Qualitative analysis	Tesch's methodology with external data analyst. Literature review as a control of identifying the similarities and uniqueness of the research findings. Guba's model for trustworthiness.
Although the approach was demonstrated reliable, students generally failed to acknowledge the potential value of self- marking in terms of feedback and as a learning opportunity, and found the process stressful. Students perceived more disadvantages than advantages in the self- marking exercise.	/	Volunteers took test under examination conditions, afterwards they returned for self- marking session: students were given back their examination script, correct responses were presented and students were asked to mark their response(s). No opportunity to discuss. An evaluation form (3- point likert scale and open questions) asked their opinion of the process.	Comparison between self- marks of the students and those of staff. Students opinions on value of the exercise, certainty of marking, and advantages and disadvantages of the process.	Quantitative analysis	Means, range, SD, correlations, Cronbach reliability coefficient alpha Percentages of responses
Both assessment forms seems to have acceptable qualities. The OA test has an acceptable curricular, instructional and criterion validity. Peer marks correlate well with tutor and final examination scores. However, learning outcomes are lower than expected. Students perceive a gap between their working in the tutorial groups and the assessment.	S1: High, moderate and low achievers S2: two groups	S1: Overall test, two topic checklists (TOC1 and TOC2), think aloud protocols and a student evaluation questionnaire and semi-structured interviews (= 3 group discussions) S2: Peer assessment (tutor, self, peer and total scores) and peer assessment questionnaire.	S1: students' results; curricular, instructional and criterion validity; and students' perceptions of the OverAll test and the relation to their learning. S2: Results and students' perceptions about peer and self assessment.	S1: Quantitative and qualitative analysis. S2: quantitative analysis.	S1: Means, SD, % S2: Reliability through generalisability scores for both groups; Pearson correlation coefficients: percentages, means and SD.

2001/06/EA	Sue Drew	Perceptions of what helps learn and develop in education.	Findings of a series of structured group sessions, which elicited students views on their learning outcomes, and what helped or hindered their development.	Qualitative data (flip charts & individual writings) from: *Small sub- group discussions *General discussions in the whole group *Individual views in writing	263 students (14 course groups).	/
2001/07/EA	Wendy Boes & Delphine Wante	Portfolios: the story of the student- in-development.	Explorative study on students' perceptions of portfolios as an instrument of guidance, assessment and evaluation.	Qualitative- interpretative research: * Research logbook * Questionnaire for students * Interview with staff * Portfolio analysis.	48 students in two Flemish institutions for teacher education.	/
2000/01/TA	Nerilee Flint	Culture club. An investigation of organisational culture.	Literature review of the findings on organizational culture. It aims at giving insight into what culture is, what the components of culture are, and how the culture of schools and universities are viewed. It is concluded that organizational culture has an incredible function and is an incredible function as it is both product and process, it is both effect and cause. It is for this reasons that it is important in the investigation of tertiary students' perceptions of the fairness of educational assessment to include an examination of the cultural influence of the course that the students are studying.			
2000/02/EA	R. A. Edelstein, H. M. Reid, R. Usatine and M. S. Wilkes.	A comparative study of measures to evaluate medical students' performances.	To assess how computer- based case simulations and standardized patient exams compare with each other and with traditional measures of medical students' performance, & to examine students' attitudes toward new and traditional evaluation modalities.	Quantitative study Students were assigned two days of performance examinations (also self- administered attitudinal survey).	155 fourth year students at University of California, School of Medicine.	/
1999/01/EA	Debra K. Meyer, & Linda F. Tusin	Pre- service teachers' perceptions of portfolio's: process versus product.	Examine students' pedagogical beliefs and their definitions of and experiences with portfolio's.	Informal surveys about portfolio's, & Motivational survey: the Patterns of Adaptive Learning Survey	20 elementary education majors in two groups, One semester through methods courses.	/

There are three areas which form the context in which students learn, and which have a strong influence on how and if they learn: (1) course organization, resources and facilities, (2) assessment and (3) learning activities and teaching. Set within this context is the student and his use of that context, relating to (1) students' self- management, (2) students motivation and needs, (3) students understanding and (4) students need for support.	/	Qualitative data from: *Small sub- group discussions *General discussions in the whole group *Individual views in writing, resulting in two products: a list of student identified learning outcomes and the students' comments about what had helped or hindered the development of the outcomes (=focus of this paper!)	Students' perceptions about what helps or hinders learning and development in education.	Qualitative analysis	Information was reviewed and sorted into comments that appeared to be similar by the author.
Students felt portfolios stimulated them to reflect and develop professionally. They felt engaged, but the construction in itself was not sufficient. Supervision was seen as desirable and necessary. Advantages were especially seen in relation to evaluation. Students thought portfolios were very time- consuming and expensive. Portfolios appear very useful when instruction and evaluation are integrated.	Two institutions for teacher education	Questionnaire for students about experiences and perceptions about portfolios. Interview with staff, about their experiences and perceptions. Making of, and analysis of the portfolios by the students.	Are the objectives of portfolio assessment achieved? What's the student's and staff's perspective?	Qualitative analysis.	Coding system
CBX and SPX had low to moderate significant correlations with each other and with traditional perf. measures. Traditional measures were inter-correlated at higher levels than with CBX and SPX. Students' perceptions of the various evaluation methods varied based on the assessment. Conclusion: perf. examinations measure different physician competency domains and multipronged (diff. formats for diff. purposes) assessment appr's are important./ Women score less on MC, but outperform men on essay!	Students' *Demographics, *Past performances *Specialty choices	Computer based examination: ten interactive and dynamic case simulations that change in response to the treatment. Standardized patient examination: eight 15-minutes SP encounters representing a broad range of clinical areas. Clinical skills survey: students had to select the best of 6 tools to assess different clinical skills.	Students' performance on CBX and SPX Students' perceptions of different evaluation methods.	Quantitative analysis.	Bivariate associations: correlations, t- test, chi- square tests of statistical significance (p<.05)
Pedagogical beliefs were stable, & didn't differentiate the two groups, although level of experience varied. Significant individual differences were found in reports about process- versus product oriented approach to teaching. Three patterns: (1) moderate (n=12), (2) product/performance (n=4), and (3) process perspective (n=4).	Two groups of pre-service teachers: student teachers (ST), & educational majors (EM).	Questionnaire, two phases: (1) early in spring (before ST began full-time ST- exp's), & (2) in late summer (as ST anticipated first year teaching and EM prepared for fall semester of coursework).	Pre-service teachers' beliefs of student portfolio's (process) and professional portfolio's (product).	Qualitative (informal survey), and Quantitative analyses (motivational survey).	Descriptive analysis Anova analysis: $F_{Performance}= 0.11$, $p<0.79$, $MSE= 0.04$, & $F_{Process}= 0.05$, $p< 0.82$, $MSE= 0.00$

1998/03/EA	Kay Sambell & Liz McDowell	The construction of the hidden curriculum: messages and meanings in the assessment of student learning.	A wide range of diverse responses by individual students to innovative or alternative assessment are described and discussed, drawing on research data.	Case studies Student and staff interviews, documentary sources and observation.	13 case studies of assessment in action in a UK university	/
1998/02/EA	Barbara Rickinson	The relationship between undergraduate student counselling and successful degree completion.	Explores students' perceptions of the level of distress they experience at two transition points: entry and completion, and effectiveness of counselling intervention.	4- year study, S1: first year entry, questionnaire survey and counselling intervention S2: final year completion, self-assessment questionnaire	S1: 15 of 44 'at risk' undergraduate first year students S2: 43 final year undergraduate students, who used counselling service.	/
1998/01/EA	Menucha Birenbaum & Rose A. Feldman	Relationships between learning patterns and attitudes towards two assessment formats.	Relation between learning pattern and attitudes towards: open- ended and multiple- choice assessment formats.	Three questionnaires: (1) attitudes toward MC and OE, (2) learning related characteristics, (3) reaction to tests questionnaire. (resp. 7-,6-, and 4-point scales)	58 students at university in Israël, 36 in arts faculty & rest in social sciences faculty	/
1997/01/CB	Noël Entwistle & Abigail Entwistle	Revision and the experience of understanding.	Exploring the nature of conceptual understanding: students' experiences of developing conceptual understanding and of explaining their understandings in the examinations.	Interview study Interviews + written comments + preliminary interviews.	I: 11 final year students (9 Psychology and 2 Zoology) WC: 11 final year psychologists PI: 15 students social and economic history 37 participants HE in Britain	/

At surface levels, there was a clear match between statements made by staff and the 'messages' received from the students. Students displayed a high level of consensus about the ways in which the new assessment method differed from their perc's of normal assessment: (1) motivation, (2) relationship with staff, and (3) different nature of learning. But they have also their own individual perspectives, which all together produce many variants of the hidden curriculum.	Assessment methods: individual research projects, assessment by oral presentation, group projects, open book exams, poster presentations, simulated professional tasks, portfolios and profiles.	Semi- structured interviews with both student groups and individuals, informal style. Also staff interviews, documentary sources and observation.	Students' (and staff's) perceptions about the assessment method used.	Qualitative analysis.	Interview transcripts and field notes.
S1: the counselling service Program helped the 15 students to develop strategies for managing their anxiety and for 'settling in' socially and academically. Everybody completed their degree program. S2: 91% thought that their academic performance had improved following counselling, 98% recorded that counselling had assisted them to deal more effectively with their problems.	S1: at risk/ non at risk of leaving S2: Pre: source of referral/ Post: number of times counselot, changes in academic performance and effect of counselot/ SCL: psychological distress.	S1: questionnaire for categorizing the students, high risk: counselling intervention program, with a review counselling session S2: self- assessment questionnaires: pre- and post- counselling, together with the SCL-90-R- questionnaire.	Students' perceptions about: S1: effectiveness of counselling intervention at first year transition point S2: impact of high levels of psychological distress on academic performance and effectiveness of counselling intervention in relation to completion.	Quantitative analysis	Percentages of students' responses.
Two patterns of relationships between the learning- related variables and the assessment attitudes: high self-concept and learning process, related to positive attitude towards OE and negative toward MC, and low test anxiety related to positive attitude OE. In addition gender differences to MC, males more favourable attitude than females.	* Gender * LR characteristics: Academic self-concept, reflective, agentic processes, & methodical study * TA: Worry and Emotionality	Three different likert- type questionnaires, all administered at once.	Attitudes towards MC and OE- examinations	Quantitative analyses	Pearson product moment correlations Canonical correlations (R_{cI}= 0.60, p< 0.001, & R_{cII}= 0.45, p<0.05)
There were marked differences in the forms of understanding which students sough. These forms differ in three ways: breadth, depth and various types of structure/ Students were experiencing their understandings as having some internal form and structure, almost as independent entities which came to control their thinking paths (= Knowledge Object !!). Conclusion: the general approach to studying can be affected by the examination format. Degree (essay) examinations may be making very different intellectual demands on students.	/	Interviews: students were asked how they knew when something was understood, what it felt like when they did not understand, and what they did to develop their understanding further.	Academic understanding through the experience of students.	Qualitative analysis	Phenomenography

1997/02/CB	Dai Hounsell	Contrasting conceptions of essay-writing.	Essay-writing as a tool for assessment and an avenue to learning. The aim of the study was to examine the students' experience of EW as a learning activity.	Empiric study Two sets of semi-structured interviews and copies of the essays.	17 second-year History undergraduates and 16 second-year Psychology undergraduates	/
1997/03/EA	Liora P. Schmelkin, Karin J. Spencer & Laura J. Labenberg	Students' perceptions of the weight faculty place on grading criteria.	Students' views of and attitudes toward grading and their perceptions of the various criteria used by faculty in the grading of courses were assessed.	Mailed survey at a private university	411 students of 625 randomly selected.	/
1997/04/EA	Kay Sambell, Liz McDowell & Sally Brown	'But is it fair?': an exploratory study of student perceptions of the consequential validity of assessment.	A qualitative study of students' interpretations, perceptions and behaviors when experiencing forms of alternative assessment, in particular its consequential validity.	Case study methodology Interviews students & staff, observations, & documentary evidence.	13 case studies of alternative assessment in practice.	/
1997/05/EA	Peter Ashworth and Philip Bannister	Guilty in whose eyes? University students' perceptions of cheating and plagiarism in academic work and assessment.	Qualitative study to discover students' perc. of cheating and plagiarism, ascertain the place of cheating and plagiarism within lived experience of students interviewed, whether perc's are wise or foolish.	Qualitative study, interviews	19 interviews as coursework towards end of masters degree unit in qualitative research interviewing. The interviewees were students not undertaking the unit.	/
1997/06/EA	Menucha Birenbaum	Assessment preferences and their relationship to learning strategies and orientations.	Inter- and intra- group differences in assessment preferences among students in 2 academic disciplines.	Motivated Learning Strategies Questionnaire (MLSQ) and the Assessment Preference Inventory (API)	172 students in 2 faculties (85 engineering/ 87 education) University in Israël	/

The most fundamental difference to emerge from the interviews lay in the students conceptions of what an essay was and what EW involved. Three qualitatively distinct conceptions were identified: essay as an argument (deep appr), a viewpoint (partly deep appr) and an arrangement (surface appr). These differences in conceptions were apparent in how an essay was defined, in the students' essay-writing procedure and in the content of their essays.	Two disciplines: History and Psychology	Two sets of semi-structured interviews. Students were invited to describe: the content of the essay, how they went about preparing it, and to draw comparisons and contrasts with other essays written for the course and to discuss various aspects of the activity of essay-writing and the course setting within which it took place.	Students' experiences of essay-writing as a learning activity against the backcloth of the two course settings.	Qualitative analysis	Phenomenography
Students believed that faculty place primary emphasis on examinations and papers with much less weight placed on a combination of other criteria, e.g. effort and improvement. These findings are contrary to the findings based on faculty.	Demographic breakdowns as: class year, full time vs part-time status, sex and school.	Likert-type questionnaire (5-points), students were asked to indicate how much weight they thought faculty put on 14 criteria in formulating their final course grades by rating each criteria on a 5-point rating scale.	Students' perceptions of the weight faculty place on grading criteria.	Quantitative analysis	Means and standard deviations. Factor analysis: a two-factor orthogonal solution was found
Students were negative when discussing trad. assessment. In contrast, when students considered new forms of assessment, their views of the educational worth of assessment changed. Altern. assessment was perceived to enable, rather than pollute, the quality of learning achieved. Trad. assessment is seen as arbitrary and irrelevant. Altern. assessment was fairer as it measured qualities, skills and competences valuable in 'real-life' contexts. Fairness related to validity!	/	Semi-structured interviews with students, interviews with staff, observations and examination of documentary evidence.	Students' perceptions of particular type of assessment.	Qualitative analysis	Initial analysis: data reduction in summary case reports, After individual case analysis, a cross-case analysis followed.
1. There is a strong moral basis to students' view, which focus on values as friendship, interpersonal trust and good learning. Punishable behaviour can be justifiable, and official approved behaviour can be dubious./ 2. The notion of plagiarism is unclear./ 3. Factors as alienation from the university due to lack of contact with staff, the impact of classes and the greater emphasis on group learning are perceived as facilitating and excusing cheating.	Interviewees were roughly categorized by sex, discipline area, maturity at entry, higher education institution and current year of study.	Interviews by students undertaking masters degree unit in qualitative research interviewing, about cheating and plagiarism. Interviewees were asked to give opinions and describe situations relevant to cheating.	Students' perceptions about cheating and plagiarism.	Qualitative analysis.	Interviews are analysed separate, common themes or issues to which students responded different, were recognized and were noted and arranged under three headings.
Individual differences in assessment preferences overshadow disciplinary group differences & Differences in assessment preferences are to a relatively large extent related to learning strategies and orientations.	Different disciplines	The subjects were administered the MSLQ for measuring their motivated learning strategies and the API for measuring their assessment preferences.	Assessment preferences Students' learning related characteristics (learning orientations & strategies)	Quantitative analysis. Mean & standard deviation of the API scales. Bivariate correlations between API and MSLQ scales.	- M educ= 3.94 > M engin= 3.66 for non conventional ass. - M teacher guided test preparation= 4.66 (highest) - Sign. Rel's between HOT and MSLQ

1997/07/EA	Paul Orsmond, Stephen Merry, et al.	A study in self-assessment: tutor and students' perceptions of performance criteria.	Implementation and evaluation of a method of student self-assessment. Theme of the importance of marking criteria.	Poster assignment on a specific aspect of nerve physiology: comparison tutor-student and student-student marking of poster work for individual marking criteria.	105 first-year undergraduate biology students working in pairs.	/	It is important to consider the individual marking criteria rather than the overall mark: overall disagreement between tutor and student, with poor students over-marking, and good students under-marking their work. Agreement between students ranged from 51 to 65%. Students thought that self-assessment made them think more, learn more, be more critical and work in a more structured way. Self-assessment is challenging, helpful and beneficial.	Student/ Tutor & Student/ Student
1997/08/CB	Paul Ramsden	The context of learning in academic departments.	How students' perceptions of teaching, assessment, and course content and structure within the natural setting of academic departments may influence how students learn.	Intensive interview study and a large scale question-naire survey.	Interviews: 57 students in six university departments Questionnaire: 2208 students in 66 departments. Lancaster University	/	Students' perceptions of assessment, choice over subject matter and methods of studying it, workload and quality of teaching in academic departments are related to the main study orientations (meaning orientation and reproducing orientation), the departmental context has also an important influence on students' attitudes towards studying.	Various departments.
1996/01/TA	Timothy F. Slater	Portfolio assessment strategies for grading first-year university physics students in the USA.	Theoretical article about the advantages of portfolio assessment, student self-reflection, showcase portfolios, checklist portfolios, open format portfolios, portfolio scoring and students' perceptions about portfolios. Students' perceptions (summary of investigations with first year university physics students): overall, students like this alternative procedure for assessment. Portfolio assessment seems to remove their perceived level of 'test anxiety'. This reduction shows up in the way students attend to class discussions, relived of their traditional vigorous note-taking duties. Students thought that they would remember what they learned much better and longer, because they internalized the material while working with it. The most negative aspect of creating portfolios is that they spend a lot of time going over the textbook or required readings. Students report that they are enjoying time spent on creating portfolios and that they believe it helps them learn physics concepts.				
1996/02/EA	Bromley H. Kniveton	Student perceptions of assessment methods.	Students were asked what qualities they perceived in continuous assessment and examinations: not their preferences, but the strengths and weaknesses of the various types of assessment.	A 9-point scale questionnaire covering a number of aspects of assessment.	292 undergraduates in Human, Environmental and Social studies departments in two universities, on a voluntary basis.	2X2 groups: *Age: Young students (n=146) vs mature students (n=146). *Gender.	Students reactions varied according to age and gender (young female and mature males (most positive), more favourable towards assessment than young males (least positive) and mature females), but the overall view was that continuous assessment should not be involved in much more than half of their grade measurement. Assessment techniques are seen as fairer and measuring a greater range of abilities. No gender difference found for MCI	Age: young/ mature students Gender
1996/03/EA	Richard G. Lomax	On becoming assessment literate: an initial look at preservice teachers' beliefs and practices.	Level of assessment literacy among (student) teachers.	Pre-instruction ass. journal, self-reflection ass. journal, copies of all ass. used by PT, semi-structured interview.	57 elementary preservice teachers before, during and after assessment course University	/	A course that viewed ass. as a part of instruction appeared to be useful to PT's. Hands-on experience needed in: (1) grading, (2) difficult parents, (3) pressures of mandated standardized tests, and (4) working with teachers with other philosophies and practices.	

Stage 1: a scientific poster was to be produced, as part of their practical work/ Stage 2: verbal instructions about the marking procedure and marking scale./ Stage 3: written instructions about requirements for poster assessment/ Stage 4: assessment exercise individual evaluation questionnaire and a poster marking form were used.	Poster marking form: over and undermarking of students in comparison to tutors and other students. Evaluation form: students' perceptions about self assessment	Quantitative analysis	Percentages (dis) agreement: S/T: 86% disagreement (56% over- and 30% under- marking), S/S: 51 to 65% agreement. Sign test: differences only significant for certain individual marking criteria.
Interviews about methods of tackling academic tasks, set as part of their normal studies. Range of tasks included problem- solving, reading, essay- writing and report writing/ Questionnaire of course perceptions with eight subscales, main categories used by students when they describe context of learning in an academic department.	Students' perceptions of assessment, teaching and courses. Attitudes and approaches to studying.	Qualitative analysis	Phenomenography
Questionnaire: 47 questions, 46 answerable on 9-point scale, not concerned with what students 'preferred', but with what they considered as characteristics of different types of assessment.	Students' perceptions of characteristics of different assessment types.	Quantitative analyses	Two way analysis of variance and where interaction, there t-test: *fairness of assessment vs exam (t=13.83, p<0.00) *ability to measure range of abilities of ass vs exam (t=19.32, p<0.000)
Follow a group of PT's from just before their assessment course through the completion of student teaching. Data sources: pre- instruction and self reflection ass, journal, ass. used while teaching, and interviews.	Extent of knowledge before and during the ass. course Extent to which knowledge is used during student teaching experience Through self reflection, adequacy of knowledge	Qualitative analyses.	7 major themes: grading, parent/ teacher confer's, informal ass, stand. tests, ass. sources, change, useful in course

1995/01/EA	Arlene Franklyn-Stokes & Stephen E. Newstead	Undergraduate cheating: who does what and why?	S1: to assess staff and students' perc's of the seriousness and frequency of diff. types of cheating. S2: to elicit self-reports by undergraduates, who gave reasons for indulging in each type of behaviour.	S1: Questionnaire, seriousness is rated on 6-point scale, frequency was measured on a percentage scale. S2: Questionnaire, reasons for 20 types of cheating, (open) questions about study motivation, ability & standards.	S1: staff (n= 20), students (n= 112) S2: 128 students from two science departments.	/
1995/02/EA	John T. E. Richardson	Mature students in higher education: II. An investigation of approaches to studying and academic performance.	To investigate whether students' responses to ASI were sign. correlated with their ages, and in addition: their entrance qualities, gender, academic persistence and performance.	Quantitative study Shortened Approaches to Studying Inventory	98 students taking the same degree course	Mature (n= 38) & non mature students (n= 60)
1992/01/EA	Menucha Birenbaum, Kikumi K. Tatsuoka & Yaffa Gutvirtz	Effects of response format on diagnostic assessment of scholastic achievement.	Comparison between parallel-stem items with identical response format (OE) and stem-equivalent items with different response format (OE vs MC) on an algebra test, using two diagnostic appr's: 'bug' analysis and rule-space analysis.	48-item diagnostic algebra test on equations to identify students' bugs in solving linear algebraic equations. Bug analysis and rule-space analysis	231 eight and ninth graders (14-15 y.) from a high school in Tel Aviv.	/
1992/02/TA	Ellie Chambers	Work-load and the quality of learning.	It is argued that 'reasonable' work-load is a pre-condition of good studying and learning. Some of the ways in which workload can be measured are discussed and in particular the methodological difficulties involved in relying on students' perceptions of it. A more rigorous method of calculating student workload is outlined.			
1991/01/EA	Noël J. Entwistle & Abigail Entwistle	Contrasting forms of understanding for degree examinations: the student experience and its implications.	Exploring through students' experiences the ways in which studying, revising and understanding reached, are affected by the assessment procedures in place.	Exploratory study Interviews and written responses.	Interviews: 13 and written responses: 11 students in their final undergraduate year.	/

S1: type of respondent (staff vs student), university (old and new institution), age (students) and behaviour type (22) S2: age, sex, reasons for studying, perceived own ability and self-standards.	S1: staff were given the questionnaires via their pigeon holes, student questionnaires were administered and collected in class. S2: student questionnaire administered at a lecture, reasons for (not) cheating.	S1: Seriousness and frequency of different cheating behaviours. S2: Reasons for cheating behaviour and relationships with other variables.	Quantitative analysis	S1: three way analysis of variance, t- test, correlations S2: frequencies, correlations
Mature/ non mature Male/ female Entrance qualifications Final degree result	Response sheets (shortened ASI) from first session and from the second session.	Students' approaches to studying	Quantitative analysis	Factor analysis, correlations coefficients, mean scores, multiple regression analysis, Mann-Whitney tests
OE vs MC format	Diagnostic algebra test, followed by bug analysis and rule-space analysis.	Effect of response format on diagnostic assessment	Quantitative analysis	Bug analysis: response vectors and mean, SD of percentage of (in)correct matched responses on two sets of OE items and OE/ MC items.
Assessment procedure: final degree test.	Interview on the students' 'experiences of revision and how they tried to develop understanding'. Written responses on 'what is understanding?'.	Students learning orientations and strategies towards the test. Students conceptions and experiences of the nature of understanding.	Qualitative analysis.	Phenomenographic analysis

ID	Authors	Title	Purpose	Sample		Results	Variables
1990/01/TA	Ross E. Traub & Katherine MacRury	Multiple-choice versus Free-response in the testing of scholastic achievement.	Review on empirical research on MC vs Free response in the testing of scholastic achievement. * Three types of studies: (1) Studies of relative difficulty and reliability of MC and FR tests, (2) studies of equivalence of traits measured by MC and FR discrete tests, and (3) studies of equivalence of traits measured by MC and FR essay tests. * Conclusions: (1) different abilities seem to be demanded by MC and FR- test items, although the differences may be inconsequential between MC and FR discrete items, the nature of the differences is not well understood, (2) research questions concerning differences in mean performance due to item format, including questions about bias for or against males or females, seem impossible to answer unambiguously because of fundamental considerations involving the scaling of scores on the two types of tests, and (3) students report that they are influenced by the expectation that a test will be in MC or FR format. In particular, more positive attitudes towards MC. Effects on learning: performance of MC test by students expecting MC is not sign. different from that of students told to expect an FR test, but students expecting an FR test performed an FR test significantly better than students told to expect an MC test.				
1990/02/EA	Menucha Birenbaum	Test anxiety components: comparison of different measures.	Investigate the structure of the text anxiety construct, & compare the psychometric properties of rating scales versus questionnaire scales of test anxiety.	172 undergraduate and graduate students enrolled in one of 5 introductory educational research methodology courses.	Six questionnaires (Likert- type)	Structure of test anxiety exists of four factors: tension, worry, bodily symptoms, and test irrelevant thinking. The patterns of relationships remained intact, when the broader context, incl. test related perceptions, was studied. Rating scales are at least as valid measures of a single trait as questionnaire scale.	Test related perceptions
1988/01/EA	Uri Zoller, & David Ben-Chaim	Interaction between examination type, anxiety state and academic achievement in college science: an action- oriented research.	Study the interaction between examination type, test anxiety and academic achievement within attempt at reducing the test anxiety and to improve performance.	83 college science students Case: 26 students, type G: freshman organic chemistry, type B: freshman biology majors.	S1: "State trait anxiety inventory form (4-point likert type) *Types of preferred examinations (likert type) and reasons S2: "Mini case study: Type G and B, state anxiety inventory.	Preferences for examinations in which emphasis is on understanding rather than 'remembering', use of relevant material is permitted and time is unlimited. Students state anxiety correlates with type of examination, with higher anxiety for females. Preferred type reduces test anxiety. Reduction of anxiety and improvement of achievement more significant for low achievers. Teachers are aware of student preferences, but don't change.	Gender Year of college Subject Midterm/ final exams Case: low/ high achievement
1987/01/EA	James W. Martin	Student perception of end-of- course evaluations.	Students' perceptions about teacher evaluations at the end of the course.	Two universities: *WIU: 313, ASU: 336 resp. economics classes at College of business. *Chairpersons: WIU= 30, ASU= 23.	*Questionnaire: nine questions, 4- point scale. *Questionnaire for chairpersons.	1. Existing procedures are adequate to evaluate the teacher, students are fair and accurate in their ratings of faculty. 2. Nobody pays much attention nor does much as a result of the outcome of the evaluation process. 3. There is little effect that a student can have on the careers of faculty.	Two universities: one major difference, faculty of WIU is unionized, ASU is not, there are individual salary differences.
1987/02/EA	Moshe Zeidner	Essay versus Multiple- Choice type classroom exams: the student's perspective.	Compare students' preferences, attitudes and perc's toward teacher-made essay versus multiple- choice type exams, & delineate construction and use of the test attitude inventory.	S1: 174 junior high school students S1': 101 seventh and eight grade students	S1: Test attitude inventory (TAI), two parts: (i) likert type rating scale, & (2) relative rating scale. S2/ S1': (Replication) Modified version of TAI, personal data inventory and self- report scholastic achievement.	Generally, more favourable student attitudes towards MC were found, compared to ET formats, on most dimensions assessed. The smallest differences between the formats were evidenced on the dimensions of thickness, perceived interest, and perceived value. Students perceive ET more appropriate for reflecting one's knowledge in the subject matter. TAI very important for educational practice.	S1 & 1: Sex S1': Social background, & perceived classroom achievement.

	Six questionnaires: (1) reactions to tests, (2) rating scales for measuring test anxiety components, (3) worry-emotionality scale, (4) general test anxiety scale, (5) effects of test anxiety on performance, and (6) test-related perceptions.	Structure of test anxiety construct; intra- and intercorrelations among test anxiety scales; & Relationships among the various test anxiety measures and other test related perceptions.	Quantitative analyses	Pearson product moment correlations, & Guttman-Lingoes Smallest Space Analysis.
	*State trait anxiety inventory form (4-point likert type) *Types of preferred examinations (likert type) and reasons Case: traditional exam or take-home examination, before exam administration of state anxiety inventory questionnaire.	Test anxiety state Preferences and reasons for preferring particular types of examination Case: change in test anxiety and achievement	Quantitative analyses.	Mean scores and SD Correlation coefficients.
	*Questionnaire, nine questions, 4-point rating scale and space for student comments. *Questionnaire, five questions, 4-point rating scale.	*Students' perceptions about end-of-course evaluation. *Chairpersons' interest in these evaluations.	Quantitative analyses.	Percentages of responses Mean values Chi square test
	S1: Test attitude inventory; semantic differential scale ratings (10 dimensions) and comparative rating scales (7 criteria). S1: Modified TAI, personal data inventory and rating of themselves with respect to scholastic achievement (5-point scale).	Students' preferences, attitudes and perceptions towards MC and ET exams.	Quantitative analyses.	Means and standard deviations (S1: M_{ET}=3.02, SD=.60/ M_{MC}=3.48, SD=.65, & S1¹: M_{ET}=4.11, SD=.96/ M_{MC}=5.44, SD=1.07) ANOVA: no sign. effects of sex, social background (SB), & sex X SB.

1984/01/EA	Irwin G. Sarason	Stress, anxiety and cognitive interference: reactions to tests.	The nature of test anxiety and its relationships to performance and cognitive interference are analysed from the standpoint of attentional processes.	S1: Test Anxiety Scale S2: TAS, Reactions To Test (RTT), Digit Symbol test & Cognitive Interference Questionnaire (CIQ) S3: RTT and experimental manipulation of instructions during tests, after CIQ.	S1: 390 introductory psychology students S2: 385 introductory psychology students S3: 180 introductory students were selected for participation in the experiment.	/ / S3: 1/3 (n= 60) contr.

Legend
TA: Theoretical article
EA: Empirical article
CB : Chapter of a book

S1: four components of test anxiety: tension, worry, test- irrelevant thinking and bodily reactions./ S2: Cognitive interference and lowered performance in evaluative situations are related to thoughts that reflect fears of failure and comparison with others rather than thoughts that are merely irrelevant to the situation./ S3: an attention- directing interpretation of anxiety and worry is supported. Such instructions provide subjects with an applicable coping strategy./ Conclusion: anxiety is, at least in evaluative situations, to a significant extent a problem of intrusive thought that interfere with task- focused thinking.	S1: / S2: Gender S3: Gender, three levels of score on Worry scale and three conditions (two experimental and one control)	S1: TAS, 91 items were group administered, 4-point scale. S2: Groups of 15 to 20 students. TAS and RTT are administered first, than Digit Symbol test and finally the subjects responded to the CIQ. S3: RTT, task : series of difficult anagrams, three conditions : attention- directing, reassurance and no additional communication; afterwards the subjects were administered the CIQ.	S1: Personal reactions to tests. S2: Information about the scales' psychometric properties and their relationships to cognitive interference. S3: tendency to worry about tests after instructions (attention- direction, or reassuring communication or no instructions)	Quantitative analysis	S1: principal components factor analysis with orthogonal varimax rotation S2: means, SD and correlations S3: 3X3X2 analysis of variance (ANOVA) design

3. STUDENTS' PERCEPTIONS ABOUT ASSESSMENT

The repertoire of assessment methods in use in higher education has expanded considerably in recent years. New assessment methods are developed and implemented in higher education, for example: self, peer, and co-assessment, portfolio assessment, performance assessment, simulations, formative assessment and OverAll assessment. The notion of "alternative assessment" is often used to denote forms of assessment which differ from the conventional assessment methods, such as multiple- choice testing and essay question exams and continuous assessment through essays and scientific reports (Sambell, McDowell & Brown, 1997). New constructivist theories and practices go together with a shift from a "test" culture to an "assessment" culture. The assessment culture, embodied in current uses of alternative assessment favors: the integration of assessment, teaching and learning; the involvement of students as active and informed participants; assessment tasks which are authentic, meaningful and engaging; assessments which mirror realistic contexts; focus on both the process and products of learning; and moves away from single test- scores towards a descriptive assessment based on a range of abilities and outcomes (Sambell et al., 1997).

In this part of the review, the literature and research on students' perceptions about assessment are reviewed. The impact of (perceived) characteristics about assessment on students' approaches to learning and vice versa, is examined and discussed. This way, an attempt is made to answer our first question of special interest to this review. Next, students' perceptions about different, new modes of assessment are presented, including: portfolio assessment; self- and peer assessment; overall assessment; simulations; and finally, more general perceptions of students about assessment are investigated. This analysis has the aim to gain insight into our second review question: "What are students' perceptions about different alternative assessment formats and methods?". Finally, the effects of students' perceptions about assessment on student learning are reviewed, and therefore an answer to our third and last question is provided.

It should be notified that there are marked differences of what "perception" means in the operational sense for various studies. Some authors define perceptions as the opinions (e.g. do you think that cheating is ethical justifiable?) that students have concerning learning and studying, cheating and plagiarism, etc. Also students' attitudes (e.g. do you find this assessment format difficult?) towards and preferences (e.g. do you prefer multiple choice test to an essay exam?) for different formats of assessment are included in the concept of "perception". Yet other researchers try to capture students' experiences (e.g. how did you handle this task?) with a particular or several assessment formats, with the word "perception".

These differences are pointed out in the text and should be taken into consideration, while interpreting the investigations, its results and its educational implications.

3.1 Assessment and Approaches to Learning

Assessment is one of the defining features of the students' approaches to learning (e.g. Marton & Säljö, 1997; Entwistle & Entwistle, 1991; Ramsden, 1997). In this part of the review, an attempt is made to gain insight into the relations between (perceived) assessment properties and students' approaches to learning and studying.

3.1.1 Approaches to Learning

When students are asked for their perceptions about learning, mainly three approaches to learning occur: (1) the surface approach to learning, (2) the deep approach to learning, and (3) the strategic or achievement approach to learning.

    a. Surface approach to learning

Surface approaches to learning describe an intention to complete the task with little personal engagement, seeing the work as an unwelcome external imposition. This intention is often associated with routine and unreflective memorisation and procedural problem solving, with restricted conceptual understanding being an inevitable outcome (Entwistle & Ramsden, 1983; Entwistle, McCune & Walker, 2001). The surface approach is related to lower quality outcomes (Trigwell & Prosser, 1991).

    b. Deep approach to learning

Deep approaches to learning, in contrast, lead from an intention to understand, to active conceptual analysis and, if carried out thoroughly, generally result in a deep level of understanding (Entwistle & Ramsden, 1983). This approach is related to high quality learning outcomes (Trigwell & Prosser, 1991). However, this deep approach is not necessarily always the "best" way, but it is the only way to understand learning materials (Entwistle, et al. 2001).

    c. Strategic or achieving approach to learning

Several students refer in their perceptions on learning to the assessment procedures they experience. Because of the pervasive evidence of the influence of assessment on learning and studying an additional category was introduced, namely the strategic or achievement approach to learning, in which the student's intention was to achieve the highest possible grades by using organized study methods and effective time- management (Entwistle & Ramsden, 1983). The strategic (or achieving) approach describes well-organized and conscientious study methods linked to achievement motivation- the determination to do well. The student relates studying to the assessment requirements in a manipulative, even cynical, manner (Entwistle, et al. 2001). The following student's comment evidences this statement:

> "*I play the examination game. The examiners play it to. ... The technique involves knowing what is going to be in the exam and how it 's going to be marked. You can acquire these techniques from sitting in the lecturer's class, getting ideas from his point of view, the form of the notes, and the books he has written- and this is separate to picking up the actual work content*" (Entwistle & Entwistle, 1991, p. 208).

3.1.2 Assessment in Relation to Students' Approaches and Vice Versa

The research on the relation between approaches to learning and assessment is dominated by the Swedish Research Group of Marton and Säljö. These two researchers (Marton & Säljö, 1997) conducted a series of studies in which they tried to influence the students' approaches to learning towards a deep approach to learning. A prerequisite for attempting to influence how people act in learning situations is to have a clear grasp of precisely how different people act. *"What is it that a person using a deep approach does differently from a person using a surface approach?"*. The learner/ reader, using a deep approach to learning, engages in a more active dialogue with the text. One of the problems with a surface approach is the lack of such an active and reflective attitude towards the text. Consequently, an obvious idea was to attempt to induce a deep approach through giving people some hints on how to go about learning (Marton & Säljö, 1997).

In his first study, Marton (1976) adopted the following procedure for influencing the approach to learning. In the experimental group, the students had to answer questions of a particular kind while reading a text. These questions were of the kind that students who use a deep approach had been found to ask themselves spontaneously during their reading. The design of this study included an immediate, as well as a delayed, retention test. This attempt to induce a deep approach through forcing people to answer questions found to be characteristic of such an approach, yielded interesting but contra- intuitive results. At one level, it was obvious that the approach taken was influenced by the treatment to which the experimental group was exposed. However, this influence was not towards a deep approach: instead, it seemed to result in a rather extreme form of surface learning. The control group, which had not been exposed to any attempts at influencing the approach taken, performed significantly better.

What happened was that the participants invented a way of answering the interspersed questions without engaging in the learning, characteristic of a deep approach. The task was transformed into a rather trivial and mechanical kind of learning, lacking the reflective elements found to signify a deep approach. What allowed the participants to transform the learning in this way was obviously the predictability of the task. They knew that they would have to answer questions of this particular kind, and this allowed them to go through the text in a way that would make it possible to comply with the demands without actually going into detail about what is said. This process can be seen as a special case of the common human experience of transformation of means into ends. The outcome of this study raises interesting questions about the conditions for changing people's approach to learning. The demand structure of the learning situation again proved to be an effective means of controlling the way in which people set about the learning task. Actually, it turned out to be too effective. The result was in reality the reverse of the original intention when setting up the experiment. The predictability of the demand structure played a central role in generating this paradoxical outcome (Marton & Säljö, 1997).

A second study (Säljö, 1975) followed. Forty university students were divided into two groups. The factor varying was the nature of the questions that the groups were asked after reading each of several chapters from an education textbook. One set of questions was designed to require a rather precise recollection of what was said in the text. In the second group, the questions were directed towards major lines of reasoning. After reading a final chapter, both groups were exposed to both kinds of questions and they were required to recall the text and summarise it in a few sentences. The results show that a clear majority of the participants reported that they attempted to adapt their learning to the demands implicit in the questions given after each successive chapter. The crucial idea of this study was that people would respond to the demands to which they were exposed. In the group that was given "factual" questions, this could be clearly seen. They reacted to the questioning through adopting a surface approach. However, in the other group, the reaction did not simply involve moving towards a deep approach. Some did, others did not. A fundamental reason underlying this was differing interpretations of what was demanded of them. Only about half the group interpreted the demands in the way intended. The other students "technified" their learning, again concentrating solely on perceived requirements. They could summarise, but could not demonstrate understanding (Marton & Säljö, 1997).

It is important to realise that the indicators of a deep approach, isolated in the research, are symptoms of a rather fundamental attitude towards what it takes to learn from texts. What happened was that some students made it an end in itself to be able to give a summary of the text after each chapter. This is thus an example of the process of technification of learning resulting in poor performance. Both studies (Marton, 1976; Säljö, 1975) illustrate that although in one sense it is easy to influence the approach people adopt when learning, in another sense, it appears very difficult. It is obviously quite easy to induce a surface approach; however, when attempting to induce a deep approach the difficulties seem quite profound. The explanation is in the interpretation (Marton & Säljö, 1997).

In a third study, Marton and Säljö (1997) asked students to recount how they had been handling their learning task and how it appeared to them. The basic methodology was that students were asked to read an article, knowing they would be asked questions on it afterwards. Besides the questions about what they remembered of its content, students were also asked questions designed to discover how they tackled this task. All the efforts, readings and re-readings, iterations and reiterations, comparisons and groupings of the researchers finally turned into an astonishingly simple picture. The students who did not get "the point" (that is, they did not understand the text as a whole) failed to do so, simply because they were not looking for it. The main difference that was found in the process of learning concerned whether the students focused on the text itself or on what the text is about: the author's intention, the main point, the conclusion to be drawn. In the latter case, the text is not considered as an aim in itself, but rather as a means of grasping something that is beyond or underlying it. It can be concluded that there was a very close relationship between process and outcome. The depth of processing was related to the quality of outcome in learning (Marton & Säljö, 1997).

The students' perceived assessment requirements seem to have a strong influence on the approach to learning a student adopts when tackling an academic task (Säljö, 1975; Marton & Säljö, 1997). Similar findings emerged from the Lancaster investigation (Ramsden, 1981) in relation to a whole series of academic tasks and to students' general attitudes towards studying. Students often explained surface approaches or negative attitudes in terms of their experiences of excessive workloads or inappropriate forms of assessment. The experience of learning is made less satisfactory by assessment methods that are perceived to be inappropriate ones. High achievement in conventional terms may mask this dissatisfaction and hide the fact that students have not understood material they have learned as completely as they might appear to have done. Inappropriate assessment procedures encourage surface approaches; yet varying the assessment questions may not be enough to evoke fully deep approaches (Ramsden, 1997).

Entwistle and Tait (1990) also found evidence for this relation between students' approaches to learning and their assessment preferences. They found that students who reported themselves as adopting surface approaches to learning preferred teaching and assessment procedures which supported that approach, whereas students reporting deep approaches preferred courses which were intellectually challenging and assessment procedures which allowed them to demonstrate their understanding. A direct consequence of this effect is that the ratings that students make of their lecturers will depend on the extent to which the lecturer's style fits what individual students prefer (Entwistle & Tait, 1995).

3.1.3   Implications for Teaching and Assessment Practice

Assessment and approaches to learning are strongly related. The (perceived) characteristics of assessment have a considerable impact on students' approaches, and vice versa. These influences can be both positive and/ or negative. The literature and research on students' perceptions of assessment in relation to the students' approaches to learning, suggest that deep approaches to learning are encouraged by assessment methods and teaching practices which aim at deep learning and conceptual understanding, rather than by trying to discourage surface approaches to learning (Trigwell & Prosser, 1991). Therefore, lectures and educational policy play an important role in creating these "deep" learning environments.

The next subsection about students' perceptions of diverse assessment formats and methods can equip us with valuable ideas, interesting tip-offs and useful information to bring this deep learning and conceptual understanding into practice.

3.2   Assessment Format and Methods

During the last decade, an immense set of alternative assessment was developed and implemented into educational practice as a result of new insights and changing theories in the field of student learning. Students are supposed to be "active, reflective, self- regulating learners". Alternative assessment practices must stimulate these activities, but do they? An attempt is made to answer this question from the students' perspective.

In this part of the review, we provide an answer to our second review question: "What are students' perceptions about new modes of assessment?" Students' perceptions about several novel assessment methods are examined and discussed. Research studies report on a variety of formats: portfolio assessment, self- and peer assessment, OverAll assessment, and simulations. Additionally, a study of Kniveton (1996) compares students' perceptions on evaluation versus continuous assessment. Based on these reviewed studies, some implications for teaching and assessment practice are given.

3.2.1 Portfolio Assessment

The overall goal of the preparation of a portfolio is for the learner to demonstrate and provide evidence that he or she has mastered a given set of learning objectives. Portfolios are more than thick folders containing student work. They are personalised, longitudinal representations of a student's own efforts and achievements. Students have to do more than memorise lecture notes and text materials because of the active creation process involved in preparing a portfolio. They must organise, synthesise and clearly describe their achievements and effectively communicate what they have learned. The primary benefit is that the integration of numerous facts to form broad and encompassing concepts is actively performed by the student instead of the instructor (Slater, 1996). Other reasons for using portfolios for assessment purposes include the impact that they have in driving student learning and their ability to measure outcomes such as professionalism (Friedman Ben-David, Davis, Harden, Howie, Ker, & Pippard, 2001). Slater (1996) gathered the findings on students' perceptions of portfolios from several studies with first-year undergraduate physics students in the USA. Qualitative data were collected through formal interviews, focus group discussions, and open- ended written surveys. Most students interviewed and surveyed report that, overall, they like this alternative procedure for assessment. Portfolio assessment seems to remove their perceived level of "test anxiety". This reduction shows up in the way students attend to class discussions, relieved of their vigorous note- taking duties. Students thought that they would remember what they were learning much better and longer than they do with the material for other classes they took, because they had internalised the material while working with it, thought about the principles, and applied physical science concepts creatively and extensively over the duration of the course. The most negative aspect of creating portfolios is that they spend a lot of time going over the textbook or required readings. Students report that they are enjoying time spent on creating portfolios and that they believe it helps them learn physics concepts (Slater, 1996).

Boes and Wante (2001) also investigated student teachers' perceptions of portfolios as an instrument for professional development, assessment and evaluation. Data were collected through portfolio- analysis, observations, informal interviews with staff and an open questionnaire for students. A sample of 48 student teachers in two Flemish institutions for teacher education was surveyed. The students felt portfolios stimulated them to reflect and demonstrated their professional development as prospective teachers. They felt engaged in the portfolio- creating- process, but portfolio construction in itself was not sufficient. Students thought that supervision, in addition, was desirable and necessary. They saw portfolios as an instrument for professional development and personal growth, but advantages were especially seen in relation to evaluation. When students did not get grades for their portfolios, much lesser efforts were made to construct the portfolio. Although portfolios are an important source of personal pride, students thought portfolios were very time- consuming and expensive. Portfolios appear very useful in learning environments in which instruction and evaluation form integrated parts (Boes & Wante, 2001).

Meyer and Tusin (1999) also examined pre-service teachers' pedagogical beliefs and their definitions of and experiences with portfolios. They investigated whether students' pedagogical beliefs were related to their definitions and experiences with portfolios. The students in this study are familiar with portfolios, as an integral part of their elementary education program. Two types of portfolios are introduced: (1) student portfolios for assessment in the classroom, and (2) professional portfolios for evaluation of teachers. It is hypothesised that the students' pedagogical beliefs and their method course experiences and field experiences are important and related influences on how pre-service teachers define and use portfolios. Whether teachers view portfolios as product or process might be an important influence on how they conceptualise and use portfolios. Pre-service teachers' pedagogical beliefs are examined in terms of their achievement goals. During one semester, a sample of 20 elementary education majors was followed through methods courses into student teaching and their first year of classroom teaching. The sample consists of two groups of pre-service teachers: students completing their final methods coursework (education majors) and students completing their student teaching (student teachers). An informal survey about portfolios and a motivational survey designed for teachers, the Patterns of Adaptive Learning Survey (PALS), were conducted. The data were collected in two phases. The results indicated that beliefs about pedagogical practices appeared stable and did not differentiate the two groups although their levels of experience varied. Student teachers had more experience with portfolios personally and in field experiences prior to student teaching. Education majors reported more experience with portfolios in methods courses. Another result is that significant individual differences were found in how the students reported their beliefs about process- versus product oriented approach to teaching. Three patterns among the pre-service teachers' self- reports of their pedagogical beliefs were found: (1) the moderate perspective (n= 12), (2) the product/performance perspective (n=

4), and (3) the process perspective (n= 4). This study shows an influence of students' pedagogical beliefs and experiences on their definitions and use of portfolios, however, the complexity of these interactions and each student's uniqueness were underestimated. It appeared that "what we thought we were teaching and modelling was not always what students were learning and perceiving" (Meyer & Tusin, 1999, p. 137).

A critical and additional comment is given by Challis (2001) who argues that portfolios have a distinct advantage over other assessment methods, as long as they are judged within their own terms, and not by trying to make them replicate other assessment processes. Portfolio assessment simply needs to be seen in terms that recognise its own strengths and its differences from other methods rather than as a replacement of any other assessment methods and procedures (Challis, 2001).

3.2.2 Self- and Peer Assessment

Self- assessment and peer assessment, as well as portfolio assessment and Overall Assessment (see 4.2.5), are typical examples of alternative assessment methods in which the progressive perspectives of the constructivist movement are central.

3.2.2.1 Self- Assessment

Orsmond and Merry (1997) implemented and evaluated a method of student self- assessment. The study concerns the importance of understanding marking criteria in self- assessment. Pairs of first-year undergraduate biology students were asked to complete a poster assignment on a specific aspect of nerve physiology. The study was designed to allow the evaluation of (1) student self versus tutor marking for individual marking criteria, and (2) student versus student marking of their poster work for individual marking criteria. In the first stage of the research, 105 students were informed that as a part of their practical work a scientific poster was to be produced in laboratory time. The overall theme was to be an aspect of nerve psychology, but the students would have the choice of the specific subject for their posters. The students were told that they had to work in pairs and they were given the precise date that the finished poster would be displayed. In a second stage, the students were given verbal instructions about the poster marking scale, and the self- marking procedure. In the third stage, the students were given written instructions about what was required for the poster assessment. The written instructions supported the previous verbal instructions. During the fourth and last stage, the assessment exercise took place. The 105 students were asked to fill in an individual evaluation questionnaire, so that students' feedback on the exercise could be obtained.

A poster marking form was used by individual students to mark each poster for five separate marking criteria. A space for students' comments was provided. An overall mark for each poster was obtained by adding the five criterion values together. Once all the posters had been marked by the students, the tutor marked the work. A comparison between the tutor and the student self-assessed mark revealed an overall disagreement of 86%, with 56% of students over-marking and 30% under-marking. It is noticeable that poor students tend to over-mark their work, whilst good students tend to under-mark. If the individual criteria are considered, than the number of students marking the same as the tutor ranged form 31% to 62%. The agreement among students' marks ranged from 51 to 65%. Students acknowledged the value of this self-marking task. They thought that self-assessment made them think more and felt that they learned more. Most of the students reported that self-assessment made them more critical of their work and they felt that they worked in a more structured way. Self-assessment is perceived as challenging, helpful and beneficial. It is concluded that marking is a subjective activity and having clear marking criteria that are known to both students and tutor allows the students to see how their marks have been obtained. It is far better to take the risk over marks than to deprive students of the opportunity of developing the important skills of making objective judgements about the quality of their own work (and that of their peers) and of generally enhancing their learning skills (Orsmond & Merry, 1997).

Mires, Friedman Ben-David, Preece and Smith (2001) undertook a pilot study to evaluate the feasibility and reliability of undergraduate medical student self-marking of degree written examinations, and to survey students' opinions regarding the process. A paper consisting of four constructed response questions was administered to 119-second year students who volunteered to take the test under examination conditions. These volunteers were asked to return for the self-marking session. Again, under examinations, 99 students who attended the self-marking session, were given back their original unmarked examination scripts. The agreed correct responses were presented via an overhead projector and students were asked to mark their responses. There was no opportunity for discussion. Prior to leaving the session, students were asked to complete an evaluation form which asked them about the value of the exercise (3-point Likert scale), certainty of marking and advantages and disadvantages of the process. In contrast to the study of Orsmond and Merry (1997), a comparison between the student's marks and the staff's marks, for each question and the examination as a whole, revealed no significant differences. Student self-marking was demonstrated to be reliable and accurate. If student marks alone had been used to determine passes, the failure rate would have been almost identical to that derived from staff marks. The students in study, however, failed to acknowledge the potential value of self- marking in terms of feedback and as a learning opportunity, and expressed uncertainty over their marks. Students perceived many more disadvantages than advantages in the self- marking exercise. Disadvantages included: finding the process stressful, feeling that they could not trust their own marking and having uncertainties on how to mark, being too concerned about passing/ failing to learn from the exercise, worrying about being accused of cheating and hence having a tendency to under-mark, having the opportunity to "cheat", finding the process tedious, considering it time consuming and feeling that the faculty were "offloading" responsibility. Advantages included the feeling of some students that it was useful to know where they had gone wrong and that feedback opportunity was useful (Mires et al., 2001).

These two studies revealed interesting but quite opposite results. The different task conditions could serve as a plausible explanation. A first task condition that differs in both studies is the clarity of the marking criteria. In the second study, for each question the agreed correct answer was presented, while in the first study, only general marking guidelines were given. These marking guidelines were not as specific and concrete as those provided by the correct answers. Another important task condition that differed, was the level of stress experienced in the situation. In the first study, the task formed a part of the practical work the students had to produce during laboratory time. This is in strong contrast to the second study, in which the task was an examination. The level of stress in this situation was high(er), because the evaluative consequences are more severe. Students' primary concern was whether they failed or passed the examination. This stressful pre- occupation with passing and, failing, is probably the reason why students could not acknowledge the potential value of the self- marking exercise for feedback purposes or as a learning opportunity.

3.2.2.2 Peer Assessment

Segers and Dochy (2001) gathered quantitative and qualitative data from a research project which focused on different quality aspects of two new assessment forms in problem- based learning: the OverAll Test (see 4.2.5) and peer assessment. Problem- based learning intends to change the learning environment towards a student- centred approach, where knowledge is a tool for effective problem analysis and problem- solving, within a social context where discussion and critical analysis are central. In the Louvain case, peer assessment was introduced for students to report on collaborative work during the tutorial meeting, and during the study period that follows these weekly meetings. Pearson correlation values indicated that peer and tutor scores are significantly interrelated. The student self- scores are, to a minor extent, related to peer and tutor scores. These findings suggest that students experience difficulties in assessing themselves. Critical analysis of their own functioning seems to be more difficult than evaluating peers. A questionnaire was developed to measure students' perceptions of the self- and peer assessment. A sample of 27 students administered the questionnaire. It was found that, on one hand, students are positive about self- and peer assessment as stimulating deep- level thinking and learning, critical thinking, and structuring the learning process in the tutorial group. On the other hand, the students have mixed feelings about being capable of assessing each other in a fair way. Most of them do not feel comfortable in doing so (Segers & Dochy, 2001).

3.2.3 OverAll Test

In the Maastricht case of Segers and Dochy (2001) their investigation, a written examination, namely the OverAll Test, was used to assess the extent to which students are able to define, analyse, and solve novel, authentic problems. It was found that the mean score on this test was between 30% and 36%, with a standard deviation from 11 to 15. This implies that the students master on average one- third of the learning goals measured. Staff perceived these results as problematic. Two topic checklists were used to assess, the extent to which the OverAll Test measures the curriculum as planned (curriculum validity) and the curriculum as implemented in practice (instructional validity). The results suggest that there is an important degree of overlap between the formal and the operational curriculum in terms of concepts studied. Additionally, there is an acceptable congruence between the assessment practices in terms of goals assessed and the formal and operational curriculum. Thus, the OverAll Test seems to have a high instructional validity. Through the analysis of think- aloud protocols of students handling real- life problems, confirmatory empirical evidence of criterion validity was found. This type of validity refers to the question of whether a student's performance on the OverAll Test has anything to do with professional problem- solving. For staff, the central question remained why students did not perform better on the OverAll Test. Therefore, students' perceptions of the learning- assessment environment were investigated. A student evaluation questionnaire was administered to 100 students. The students' negative answer to the statement "the way of working in the tutorial group fits the way of questioning in the OverAll Test" particularly struck the staff as contradictory. Although empirical evidence of curriculum validity was found, students did not perceive a match between the processes in the tutorial group and the way of questioning in the OverAll Test. Staff regarded this perception as a serious issue, particularly because working on problems is the main process within problem- based learning environments. In order to gain more insight into these results, semi-structured interviews were done in four groups (n total = 33). The students indicated that the other assessment instruments of the curriculum mainly measured the reproduction of knowledge. Students felt that for the OverAll Test, they had to do more; they had to build knowledge instead of merely reproducing it. The tutorial group was perceived as not effectively preparing students for the skills they need for the OverAll Test. Too many times, working in the tutorial groups was perceived as running from one problem to another, without really discussing the analysis and the solution of the problem, based on what was found in the literature. The students also indicated that they had problems with the novelty of the problems. During the tutorials, new examples, with slight variations to the starting problem are seldom discussed. The students suggested more profound discussions in the tutorial groups, and that analysing problems should be done in a more flexible way. In one of the modules, a novel case was structurally implemented and discussed in the tutorial groups based on a set of questions similar to the OverAll Test questions. Students valued this procedure, and felt the need to do this exercise in flexible problem analysis, structurally in all modules (Segers & Dochy, 2001).

From both cases, the Louvain and the Maastricht case, it can be concluded that there is a mismatch between the formal learning environment as planned by the teachers and the actual learning environment as perceived by the students. Students' perceptions of the learning- assessment environment, based on former learning experiences and their recent experiences, have an important influence on their learning strategies and affect the quality of their learning outcomes. Therefore, they are a valid input for understanding why promises are not fulfilled. Moreover, looking for students' perceptions of the learning- assessment environment seems to be a valid method to show teachers ways to improve the learning-assessment environment (Segers & Dochy, 2001).

3.2.4 Simulation

Edelstein, Reid, Usatine and Wilkes (2000) conducted a study to assess how computer- based case simulations (CBX) and standardised patient exams (SPX) compare with each other and with traditional measures of medical students' performance. Both SPX and CBX allow students to experience realistic problems and demonstrate the ability to make clinical judgements without the risk of harm to actual patients. The object of the study was to evaluate the experiences of an entire senior medical school class as they took both traditional standardised examinations and new performance examinations. In a quantitative study, 155 fourth- year students of the School of Medicine at the University of California, were assigned two days of performance examinations. After completing the examinations, the students filled in a paper- and- pencil questionnaire on clinical skills. The examination scores were linked to the survey and correlated with archival student data, including traditional performance indicators. It was found that the CBX and the SPX had low to moderate statistically significant correlations with each other and with traditional measures of performance. Traditional measures inter-correlated at higher levels than with CBX or SPX. Students' perceptions of the various types varied based on the assessment. Students' rankings of relative merits of the examinations in assessing different physician attributes evidenced that performance examinations measure different physician competency domains. Students individually and in subgroups do not perform the same on all tests, and they express sensitivity to the need for different purposes. The use of multiple evaluation tools allows finer gradations in individual assessment. A multidimensional approach to evaluation is the most prudent (Edelstein et al., 2000).

3.2.5 Evaluation Versus Continuous Assessment

In his study, Kniveton (1996) asked students what qualities they perceived in continuous assessment and examinations. The important question is not what students "like", but what they feel are strengths and weaknesses of various types of assessment. Subjecting the student to an assessment procedure that the student can react to positively may well be an important contributor to a student's success, and the use to which a particular assessment technique can be put will to some extent depend on the student's perceptions on it. A questionnaire, with 47 questions of which 46 are answerable on a 9-point scale, concerning what students considered characteristics of the different types of assessment, was used. This instrument was administered to 292 undergraduates in human, environmental and social studies departments in 2 universities. It is the purpose of the research to examine and compare the perceptions of students taking a number of degrees, giving equal weight to the variables of age and gender. The overall view of the students was that continuous assessment should not be involved in much more than half of their grade measurement. Although assessment techniques are seen as fairer and measuring a range of abilities, this finding does not indicate an overwhelming endorsement of continuous assessment, nor does it indicate a total rejection of the idea of examinations. There are a number of sub- group differences found. First, there are a number of aspects of assessment where there is an interaction between gender and age. Mature males and younger females tend to regard continuous assessment as having many advantages over examinations. Younger male and mature female students are far less positive about continuous assessment. At a second level, there are a number of aspects of assessment where mature male students more than other groups feel that aspects of continuous assessment are extremely positive. On average mature males want the most continuous assessment and younger males the least (Kniveton, 1996).

3.2.6 General Perceptions About Assessment

A series of studies do not focus on students' perceptions about specific modes of assessment but more generally investigate students' perceptions about assessment. The study of Drew (2001) illustrates students' general perceptions about the value and purpose of assessment. Within the context of new modes of assessment, the Northumbria Assessment studies are often cited. In these studies, different aspects of perceptions of students about new modes of assessment are elaborated upon: the consequential validity of alternative assessment and its (perceived) fairness, but also the relations between teacher's messages and student's meanings in assessment, and the hidden curriculum are investigated.

3.2.6.1 What Helps Students Learn and Develop in Education

Drew (2001) describes the findings of a series of structured group sessions, which elicited students' views on their learning outcomes, and what helped or hindered their development.

The process of amended session consisted of: (1) small sub-group discussions, (2) general discussions in the whole group, and (3) students' individual views in writing. The amended session was run with 14 course groups in Sheffield Hallam University, with a total of 263 students. Each session generated an amount of qualitative data, in the form of student-generated flip charts and individually written views. The students' comments about what helped or hindered the development of their learning outcomes are the focus of the researcher. The findings suggest that there are three areas (i.e. three contextual factors) that, together, comprise the context in which students learn, and which have a strong influence on how and if they learn: (1) course organisation, resources and facilities, (2) assessment, and (3) learning activities and teaching. Set within this context is the student and his use of that context (i.e. four student-centred factors), relating to (a) students' self-management, (b) students motivation and needs, (c) students understanding and (d) students need for support.

Drew (2001) found following results on the four student-centred factors: (a) Students' self-management. Autonomy and responsibility for their learning were themes emerging through students' comments. Students acknowledge the importance of operating autonomously. They liked "to be treated like adults". (b) Students' motivation and needs. The students felt it was important for allowances to be made for their individual needs, but considered that lectures often assumed their needs were identical. Students thought it was dangerous to assume that all students on a course shared interests and aspirations. Subjects needed to be pitched at their level. (c) Students' understanding. The students wanted to grasp principles and concepts, rather than detail, saw dangers in merely memorising information and thought that understanding the aims for a subject helped them to handle it. Students saw reflection as valuable and important for understanding. (d) Students' need for support. Personal, but especially academic needs for support were mentioned. Students wanted it to reduce uncertainty and anxiety, and saw support as taking a variety of forms, for example: clear structures, guidance and personal contact (Drew, 2001).

Within the context of "assessment", the second contextual factor and the focus of this review, these student-centred factors occur as follows: students valued self-management and, generally, examinations were seen as less supportive of its development. Deadlines were not seen in themselves as unhelpful. They developed self-discipline, the ability to work under pressure and increased determination, but they were also seen as indicating when to work, rather than when work was to be completed. Assessment, seen by the students as a powerful motivator, was regarded as a major vehicle for learning. However, a heavy workload could affect the depth at which they studied and, in some courses, students thought it should be lessened so that "work doesn't just wash over students". In order to help them learn, students wanted to know what was expected- clear briefs and clear assessment criteria. Students closely linked the provision of feedback with support. Effective feedback was critical to "build self confidence, help us evaluate ourselves" and students wanted more of it. Students preferred 1:1 tutorials as a method to provide effective feedback, but they knew that staff pressures made this difficult. They disliked one-line comments and saw typed feedback sheets as excellent (Drew, 2001).

3.2.6.2 But Is It Fair: Consequential Validity of Alternative Assessment

Sambell, McDowell and Brown (1997) conducted a qualitative study of students' interpretations, perceptions and behaviours when experiencing forms of alternative assessment, in particular its consequential validity (i.e. the effects of assessment on learning and teaching). The "Impact of Assessment" project has employed the case study methodology. Data were gathered from thirteen case studies of alternative assessment in practice. The methods for collecting these data included interviewing both staff and students, observation and examination of documentary evidence, but the emphasis was on semi-structured (group) interviews with students. A staged approach to interviewing was used, so that respondents' perceptions and approaches were explored over the period of the assessment, from the initial assessment briefings at the beginning of a unit of learning to post-assessment sessions. Initial analysis of the data was conducted at the level of the case, which resulted in summary case reports. Individual case analysis was followed by cross- case analysis (Sambell et al., 1997).

3.2.6.2.1 Effects of Student Perceptions of Assessment on the Process of Learning

Broadly speaking, it was discovered that students often reacted very negatively when they discussed what they regarded as "normal" or traditional assessment. One of the most commonly voiced complaints focused upon the perceived impact of traditional assessment on the quality of learning achieved. Many students expressed the opinion that normal assessment methods had a severely detrimental effect on the learning process. Exams had little to do with the more challenging task of trying to make sense and understand their subject. By contrast, when students considered new forms of assessment, their views of the educational worth of assessment changed, often quite dramatically. Alternative assessment was perceived to enable, rather than pollute, the quality of learning achieved. Many made the point that for alternative assessment they were channelling their efforts into trying to understand, rather than simply memorise or routinely document, the material being studied. Yet, although all the students interviewed felt that alternative assessment implied a high- quality level of learning, some recognised that there was a gap between their perceptions of the type of learning being demanded and their own action. Several claimed they simply did not have the time to invest in this level of learning and some freely admitted they did not have the personal motivation (Sambell et al., 1997).

3.2.6.2.2 Perceptions of Authenticity in Assessment

Many students perceived traditional assessment tasks as arbitrary and irrelevant. This did not make for effective learning, because they only aimed to learn for the purposes of the particular assessment, with no intention of maintaining the knowledge for the long- term. Normal assessment was seen as something they had to endure, not because it was interesting or meaningful in any sense other than it allowed them to accrue marks, an unavoidable evil. Normal assessment activities are described in terms of routine, dull artificial behaviour. Traditional assessment is believed to be inappropriate as a measure, because it appeared, simply to measure your memory, or in case of essay- writing tasks, to measure your ability to marshal lists of facts and details. Students repeatedly voiced the belief that the example of alternative assessment under scrutiny was fairer than traditional assessment, because by contrast, it appeared to measure qualities, skills and competencies that would be valuable in contexts other than the immediate context of assessment. In some of the cases, the novelty of the assessment method lay in the lecturer's attempt to produce an activity that would simulate a real life context, so students would clearly perceive the relevance of their academic work to broader situations outside academia. This strategy was effective and the students involved highly valued these more authentic ways of working. Alternative assessment enabled students to show the extent of their learning and allowed them to articulate more effectively and precisely what they had assimilated throughout the learning program (Sambell et al., 1997).

3.2.6.2.3 Student Perceptions of the Fairness of Assessment

The issue of fairness, from the student perspective, is a fundamental aspect of assessment, the crucial importance of which is often overlooked or oversimplified from the staff perspective. To students, the concept of fairness frequently embraces more than simply the possibility of cheating: it is an extremely complex and sophisticated concept that students use to articulate their perceptions of an assessment mechanism, and it relates closely to our notions of validity. Students repeatedly expressed the view that traditional assessment is an inaccurate measure of learning. Many made the point that end- point summative assessments, particularly examinations that took place only on one day, were actually considerably down to luck, rather than accurately assessing present performance. Often students expressed concern that it was too easy to leave out large portions of the course material, when writing essays or taking exams, and still do well in terms of marks. Many students felt unable to exercise any degree of control within the context of the assessment of their own learning. Assessment was done to them, rather than something in which they could play an active role. In some cases, students believed that what exams actually measured was the quality of their lecturer's notes and handouts. Other reservations that students blanketed under the banner of "unfairness", included whether you were fortunate enough to have a lot of practice in any particular assessment technique in comparison with your peers (Sambell et al., 1997). When discussing alternative assessment, many students believed that success more fairly depended on consistent application and hard work, not a last minute burst of effort or sheer luck. Students use the concept of fairness to talk about whether, from their viewpoint, the assessment method in question rewards, that is, looks like it is going to attach marks to, the time and effort they have invested in what they perceive to be meaningful learning. Alternative assessment was fair because it was perceived as rewarding those who consistently make the effort to learn rather than those who rely on cramming or a last- minute effort. In addition, students often claimed that alternative assessment represents a marked improvement: firstly in terms of the quality of the feedback students expected to receive, and secondly, in terms of successfully communicating staff expectations. Many felt that openness and clarity were fundamental requirements of a fair and valid assessment system. There were some concerns about the reliability of self and peer assessment, even though students valued the activity (Sambell et al., 1997).

3.2.6.3 The Hidden Curriculum: Messages and Meanings in Assessment

Sambell and McDowell (1998) focus upon the similarities and variations in students' perspectives on assessment, based on two levels of data analysis. At the first level, the whole dataset was used to examine the alignment between the lecturers' stated intentions for the innovation in assessment and the "messages" students received about what they should be learning and how they should go about it, in order to fulfil their perceptions of the new assessment requirements. This level revealed that, at surface levels, there was a clear match between statements made by staff and the "messages" received by students. Several themes emerged, indicating shifts in students' characterizations of assessment. First, students consistently expressed views that the new assessment motivated them to work in different ways. Second, that the new assessment was based upon a fundamentally different relationship between staff and students, and third, that the new assessment embodied a different view of the nature of learning. At the second stage of analysis, data were closely investigated on the level of the individual, to look for contradictory evidence, or ways in which, in practice, students expressed views of assessment which did not match these characterizations, and in which the surface- level close alignment of formal and hidden curriculum was disrupted in some way. It was found that students have their individual perspectives, all of which come together to produce many variants on a hidden curriculum. Students' motivations and orientations to study influence the ways in which they perceive and act upon messages about assessment. Students' views of the nature of academic learning influence the kinds of meaning they find in assessment tasks and whether they adopt an approach to learning likely to lead to understanding or go through the motions of changing their approach (Sambell & McDowell, 1998). Students' characterizations of assessment, based on previous experience, especially in relation to conventional exams, also, strongly influence their approach to different assessment methods. In an important sense, this research makes assessment problematic, because it suggests that students, as individuals, actively construct their own versions of the hidden curriculum from their experiences with and characterizations of assessment. This means that the outcomes of assessment as "lived" by students are never entirely predictable, and the quest for a "perfect" system of assessment is, in one sense, doomed from the outset (Sambell & McDowell, 1998).

3.2.7 Implications for Teaching and Assessment Practice

Previous educational research on students' perceptions about conventional evaluation and assessment practices, namely multiple choice and essay typed examinations, evidence that students perceive the multiple choice format as more favourable than constructed response/ essay items on following dimensions: perceived difficulty, anxiety, complexity, success expectancy and feeling at ease (Zeidner, 1987). Within these groups of students, some remarkable differences are found. Students with good learning skills and students with low test anxiety rates, both seem to favour the essay type exams (Birenbaum & Feldman, 1998). This type of examination goes together with deep(er) approaches to learning than multiple-choice formats (Entwistle & Entwistle, 1991).

When compared to alternative assessment, these perceptions about conventional assessment formats seem to contradict strongly the students' more favourable perceptions towards alternative methods. Overall, learners think positive about new assessment strategies, such as portfolio assessment, peer assessment, simulations and continuous assessment methods.

Although students acknowledge the advantages of these methods, some of the students' comments put this overall positive image of alternative assessment methods into perspective. Different examination or task conditions can interfere. For example, "reasonable" work- load is a precondition of good studying and learning (Chambers, 1992). Sometimes, a mismatch was found between the formal curriculum as intended by the educator and the actual learning environment as perceived by the students. Furthermore, different assessment methods seem to assess various skills and competencies. It is important to value each assessment method, within the learning environment for which it is intended, and taking its purposes and skills to be assessed into consideration, as well as the cost- benefit profile of each different mode. For example, is it appropriate to adapt the assessment automatically to (each of) the student's preferences? Regarding your instruction and your assessment method as integrated parts, do they have the same or compatible purposes? How about the time investment for the students and/ or the teacher's time investment on this particular assessment type?

In addition, methodological issues, like a poor operational implementation of the assessment mode or format, can give rise to biased results about students' perceptions on several types of assessment. Any assessment done poorly will result in poor results. Therefore, it is important to consider the research methodological design when interpreting the findings, and certainly when assessing, evaluating and changing teaching practices. Further research is needed to verify and consolidate the results of these investigations.

3.3 Effects of Perceptions about Assessment on Student Learning

As we have already shown, students' perceptions about assessment have an important influence on students' approaches to learning. However, are those the only influences? We studied the effects of students' perceptions about assessment on their learning, and thus be in a position to provide an answer to our third and final review question.

3.3.1 Test Anxiety

Test anxiety can have severe consequences for the student's learning outcomes. In this section, the intrusive thoughts and concerns of the student with(out) test anxiety are investigated.

3.3.1.1 Nature of Test Anxiety

Sarason (1984) analysed the nature of test anxiety and its relationships to performance and cognitive interference from the standpoint of attentional processes. The situations to which a person reacts with anxiety may be either actual or perceived. The most adaptive response to stress is task- oriented thinking, which directs the individual's attention to the task at hand. The task- oriented person is able to set aside unproductive worries and preoccupations. The self- preoccupied person, on the other hand, becomes absorbed in the implications and consequences of failure to meet situational challenges. The anxious person's negative self- appraisals are not only unpleasant to experience, they also have undesirable effects on performance because they are self- preoccupying and detract from task concentration. Sarason (1984) conducted three studies, concerning an instrument, Reaction To Tests (RTT), designed to assess multiple components of a person's reactions to tests, to correlate those components with intellective performance and cognitive interference, and to attempt experimentally influence these relationships. In the first study, a pool of items (Test Anxiety Scale) dealing with personal reactions to tests was constructed and administered to 390 introductory psychology students. The findings of this study indicate the existence of four discriminable components of test anxiety: Tension, Worry, Test- Irrelevant Thinking, and Bodily Reactions. Based on of these findings, a new instrument, the Reactions To Tests questionnaire, was developed and administered to 385 psychology students. This second study was conducted to obtain information about the scales' psychometric properties and to determine their relationships to cognitive interference. The subjects first filled in the RTT and the TAS, then they were given a difficult version of the Digit Symbol Test and immediately after this, they responded to the Cognitive Interference Questionnaire (CIQ). It was found that the Worry scale related negatively to performance and related positively to cognitive interference and thus that test anxiety is best conceptualised in terms of worrisome, self- preoccupying thoughts that interfere with task performance. The third study was carried out in an effort to compare groups that differ in the tendency to worry about tests after they have received either (1) instructions directing them to attend completely to the task on which they will perform, or (2) a reassuring communication prior to performing the task. From a group of 612 students who responded to the RTT, 180 introductory psychology students were selected for participation in the experiment. The findings show that reassuring instructions have different effects for subjects who score high, moderate and low on the Worry scale, especially "worriers" seem to have advantage of the reassuring instructions prior to the performance task. There is a detrimental effect of reassurance on the students who score low on the Worry scale. This may be due to the student's interpretation of the reassuring communication as the task being to lightly. This might lower their motivational level and as a consequence, their performance. The attention- directing condition seems to have all the advantages that reassurance has for high Worry scale scorers with none of the disadvantages. The performance levels of all groups receiving these instructions, were high. The attention- directing instructions seemed to provide students with an applicable coping strategy. The results of the present studies suggest, at least in evaluation situations, anxiety is to a significant extent, a problem of intrusive, interfering thoughts that diminish attention to and efficient execution of the task. Under neutral conditions, high and low test- anxious subjects perform comparably. The study evidenced that it is possible to influence these thoughts experimentally. People who are prone to worry in evaluative situations benefit simply from their attention being called to the importance of maintaining a task focus. Reassurance, calming statements geared to reduce the general feeling of upset that people experience in threatening situations, can be counterproductive, especially for students with low and moderate anxiety scores (Sarason, 1984).

3.3.1.2 Test Anxiety in Relation to Assessment Type and Academic Achievement

The main objective of Zoller and Ben- Chaim (1988) was to study the interaction between examination type, test anxiety and academic achievement within an attempt at reducing the test anxiety of students in college science- through the use of those kinds of examinations preferred by them- and thus, hopefully, to improve their performance accordingly. The Stait Trait Anxiety Inventory (STAI; Spielberger Gorsuch, & Lushene, 1970) and the Type Of Preferred Examinations (TOPE) questionnaire were administered to 83 college science students. In this latter questionnaire, students' preferences and the reasons accompanying these preferences were assessed for several traditional and non- traditional examinations. The most preferred types of examinations are those in which the use of any supporting material (i.e. notes, textbooks, tables) during the examination is permitted, and the time duration of the exam is practically unlimited, in particular: (1) take home exam, any material may be used, and (2) written exam in class, time unlimited, any supporting material is allowed. Students emphasise the importance of the examination as a learning device, to enhance understanding, thoroughness and analysis, rather than superficial rote learning and memorisation. As expected, it was found that students believe that compared with the conservative paper- and- pencil- type examinations, the written examinations with open books either in class or at home, reduce tension and anxiety, improve performance, and are therefore perceived to be preferable. Students also claimed to have difficulty expressing themselves orally. Furthermore, it is significant that most of the science students, regardless of their year of study, are convinced and strongly believe that the type of the final examination crucially affects their final grade in the course. It also appeared that students' state of anxiety in the finals is higher than in the midterms for all four science courses. Finally, there is an important gender effect: the state anxiety level of female science students under test situations seems to be consistently higher, compared with that of male science students. If these findings are compared with a preliminary survey of the college science professors concerning the issue of examinations, a remarkable result is attained. Although teachers know precisely the types of examinations preferred by the students, each professor continues persistently, to give the students the same one type of examination, which he prefers, or considers to be the most appropriate for his needs, regardless of the students' preferences. Moreover, there exists no tendency among the science professors to divert from their "pat" examination type or to modify even slightly (Zoller & Ben-Chaim, 1988).

In a mini case study, Zoller and Ben-Chaim (1988) compared the traditional class examination (written exam in class, time limited, no supporting material is allowed) with the non- traditional take home exam (any material may be used) concerning the interaction between examination type, (test) anxiety state, and academic performance. The examination was divided in two equivalent parts that were administered in class, and as a take home exam a day apart. Each exam was accompanied by the administration of the State Anxiety Inventory questionnaire just before the initiation of the exam itself. A negative correlation between test anxiety and academic achievements was found. The lower the level of state anxiety is, the higher the students' achievements are, the difference being statistically significant. In particular, the group of low achievers gained significantly more in academic achievement in the "take home" exam, compared with the group of high achievers, whereas the level of state (test) anxiety of the low achievers decreased considerable. There was no gain in achievement of the group of high achievers in the take home exam nor a significant change in their state anxiety (Zoller & Ben-Chaim, 1988).

3.3.2 Student Counselling

Student counselling is often claimed to be a potential method to cope with high levels of distress. But is it? What are students' perspectives? Rickinson (1998) examined students' perceptions about their experienced distress and about the effects of student counselling on this distress, and related them to the student's degree completion. The study explores undergraduate students' perceptions of the level of distress they experience at two important transition points: first year entry and final year completion, and the impact of this distress on their academic performance. In addition, the effectiveness of counselling intervention in ameliorating this distress, and in improving students' capacity to complete their degree programs successfully, is discussed. During a four- year study, the relationship between undergraduate student counselling and successful degree completion is investigated. First, the research examined the effectiveness of counselling intervention at the first year transition point in relation to student retention and subsequent completion. Students were categorised into risk groups according to their level of commitment/ risk of leaving. Of the 44 students identified as "high risk", only 15 students accepted counselling intervention. At their initial counselling intervention all 15 students were assessed as having significant difficulty with academic and social integration into the university. All students attended the full workshop program and reported that the workshops had helped them to develop strategies for managing their anxiety and for "settling in" socially and academically. Of the 15 students, 11 achieved an upper second class degree, three achieved a lower second class degree and one achieved a third class honours degree. Second, the study focused on final year students, investigating both the impact of high levels of psychological distress on their academic performance and the effectiveness of counselling intervention in relation to degree completion. For final year students a self- completion questionnaire was chosen as the most practical method of assessing the perceived effect of students' problems on their academic performance both prior to, and following, counselling intervention. The self- completion questionnaires were administered to a selected sample of 43 undergraduates who used the counselling service, together with the SCL- 90- R, a psychometric instrument. Of this sample, 30 students had self- referred and 13 were referred via their tutor or doctor. Almost all students (n= 41) perceived their academic performance as having been affected by their problems prior to the counselling intervention. Following counselling, students recorded their perception of the degree of change in their academic performance and the degree to which they felt better able to deal with their problems. Of the 43 students, 39 thought that their academic performance had improved following counselling and 42 students recorded that counselling had assisted them to deal more effectively with their problems. All 43 students completed their degree programs successfully. This study highlights the educational implications of high levels of psychological distress for undergraduate students. The university learning process, by providing the stimulus of new knowledge and experience, challenges students' existing level of development. To take full advantage of this developmental opportunity, students need to tolerate the temporary loss of balance. Counselling intervention was shown to be effective in facilitating student retention and completion. Counselling assisted students at risk of leaving, to adjust to the new social and academic demands of the university environment. Subsequently, these students progressed to successful degree completion. At the second transition point, the results strongly suggest that counselling intervention was instrumental in reducing the level of psychological distress of the final year students (Rickinson, 1998).

3.3.3 Cheating and Plagiarism

Do students' perceptions about cheating and plagiarism have important consequences for students' cheating behaviour and student learning? We tried to find an answer.

Students' Perceptions about New Modes of Assessment 215

Ashworth and Bannister (1997) conducted a qualitative study to discover students' perceptions of cheating and plagiarism in higher education. The study tries to elicit how cheating and plagiarism appear from the perspective of the student. Nineteen interviews were carried out as a coursework toward the end of a semester-long Master's degree unit in qualitative research interviewing. The work was undertaken by the course members, who interviewed one student and completed a full analysis and report on that one interview.

Further analysis was done by the researchers. A first important result is that there is a strong moral basis in students' views on cheating and plagiarism, which focus on values as friendship, interpersonal trust and good learning. Practices that have a detrimental effect on other students are particularly serious and reprehensible. The ethic of peer loyalty is a dominant one. It appears that the "official" university view of cheating is not always appropriate. This means that some punishable behaviour can be regarded as justifiable and some officially approved behaviour can be felt to be dubious. Another interesting finding is that the notion of plagiarism is regarded as extremely unclear. Students are unsure about precisely what should and should not be assigned to this category. Doubt over what is "officially" permitted and what is punishable, appeared to have caused considerable anxiety. Some students have a fear that they might plagiarise unwittingly in writing what they genuinely take to be their own ideas, that plagiarism might occur by accident. Controversy, cheating which is extensive and intended and leading to substantial gain is seen as the most serious. In this respect, examination cheating is seen as more serious than coursework cheating. Finally, the study revealed that factors such as alienation from the university due to lack of contact with staff, the impact of large classes, and the greater emphasis on group learning are perceived by students themselves as facilitating and sometimes excusing cheating. For example, different forms of assessment offer different opportunities for cheating. The informal context in which coursework exercises are completed means there is an ample scope to cheat through collusion and plagiarism, in contrast to the controlled, invigilated environment of unseen examinations. This study reveals the importance of understanding the students' perspective on cheating and plagiarism; this knowledge can significantly assist academics in their efforts to communicate appropriate norms. Without a basic commitment on the part of the students to the academic life of the institution, there is no moral constraint on cheating or plagiarism (Ashworth & Bannister, 1997).

Franklyn-Stokes and Newstead (1995) conducted also two studies on undergraduate cheating. The first study was designed to assess staff and students' perceptions of the seriousness and frequency of different types of cheating. Because of the sensitivity of the topic under investigation, it was decided not to report their own cheating, instead they were asked to estimate how frequently they thought cheating occurred in their year group. A sample of 112 second- year students and 20 staff members were administered a questionnaire, who asked them to rate the frequency and seriousness of each type of a set of cheating behaviours. An inverse relationship between perceived frequency and seriousness of cheating behaviour was found: the types of cheating behaviour rated as most serious, were also rated as the least frequent. Cheating behaviour that was examination- related, was rated most serious and least frequent, while coursework- related cheating behaviours were rated least serious and occurred most frequent. There were considerable staff/ student differences in the seriousness and frequency ratings. There was no behaviour that students rated significantly more serious than did staff. The differences for frequency were even more marked. Students rated every type of behaviour as occurring more frequently and this difference was significant for 19 out of the 22 types of cheating behaviours in the questionnaire. In addition, an important age effect was found for students perceptions of cheating. The 25+ students rated cheating significant more serious and as occurring significantly less frequently, than did younger peers. There were no significant gender differences. In their second study, Franklyn-Stokes and Newstead (1995) utilised this set of cheating behaviours to elicit undergraduates' self- reports and reasons for (or not) indulging in each type of behaviour. The questionnaire required subjects to say whether they had indulged in each type of behaviour as an undergraduate. Then, they were asked to select a raison for indulging (or not) this type of cheating. Finally, in an open question, the students were asked to give the main raison why they were studying for a degree. The questionnaire was completed by 128 students from two science departments in the same university. It was found that the overall occurrence of cheating largely corroborated the findings from the first study regarding the frequency of occurrence of each type of behaviour. There was no significant gender effect. On the contrary, the difference in reported cheating by age was significant. The 18- 20 year- olds reported an average cheating rate of 30%, the 21-24 year- olds one of 36% and the 25+ students also reported an average cheating rate of 30%. The reasons for cheating and for not cheating varied to a considerable extent in relation to the type of behaviour. There was no relationship between the reason students gave for studying for a degree and the amount of cheating they admitted to. These two studies suggest that more than half of the students are involved in a range of cheating behaviours, including: allowing coursework to be copied, paraphrasing without acknowledgement, altering and inventing data, increasing marks when students mark their own work, copying another' work, fabricating references and plagiarism from text. Other important results are that cheating occurs more in relation to coursework than with examinations and that although mature students perceive cheating as less frequent and more serious, their self- reported frequency of occurrence was the same as that for 18-20 year- olds. As to the reasons why students cheat, the principal ones are time pressure and desire to increase the mark. The most common reason for not cheating, are that it was unnecessary or that it would have been dishonest. Clearly, cheating may occur more frequently than staff seem to be aware of, and it is not seen as seriously by students as it is by staff (Franklyn-Stokes & Newstead, 1995).

3.3.4 Implications for Teaching and Assessment Practice

Students' perceptions about assessment seem to have an important influence on student learning. Test anxiety and its accompanying intrusive thoughts and concerns about possible consequences of the test, have a detrimental influence on students' learning outcomes. Simple attention-directing instructions from the teacher, can equip the test anxious student with an appropriate coping strategy. In addition, the assessment type can reduce the level of test anxiety. Furthermore, students thought that counselling positively changed their academic performance and they felt they were better able to deal more effectively with their problems. Students' perceptions about cheating and plagiarism do seem to have an influence on student learning. For example, the higher the perceived seriousness of the cheating behaviour, the lower the frequency and the lower the perceived seriousness, the more frequent the cheating behaviour was. Students' perceptions have in this respect an important additional value when considering teaching and assessment practices.

4. METHODOLOGICAL REFLECTIONS

Traditionally research with regard to human learning, was done from a first order perspective. This research emphasised the description of different aspects of reality; reality per se. Research on students' perceptions turned the attention to the learner and certain aspects of his/her world. This approach is not directed to the reality as it is, but more to how people view and experience reality. It is called a second- order perspective. The accent of this second- order perspective is on understanding and not on explanation (Van Rossum & Schenk, 1984). Both qualitative and quantitative research has been conducted to reveal this second- order perspective. Especially the quantitative research concerning students' perceptions about assessment had a clear majority, 23 out of 36 studies were solely analysed by quantitative methods. Only 11 investigations, not yet one third of the reviewed studies, have been analysed qualitatively and two reviewed studies are both analysed quantitatively and qualitatively. Very popular methods for data collection within the quantitative research are the Likert type questionnaires (n= 35) and inventories (n= 7), for example: the Reaction To Test questionnaire (RTT) (Birenbaum & Feldman, 1998; Sarason, 1984), Clinical Skill Survey (Edelstein et al., 2000), Assessment Preference Inventory (API) and the Motivated Learning Strategies Questionnaire (MLSQ) (Birenbaum, 1997). Only a relatively small number of surveys (n= 7) was done in response to a particular assessment task or, in response to a test or examination. Most other studies ask for students' perceptions in more general terms, not related to the experiences with a specific assessment task. The most frequent used methods for data collection within the qualitative research, were open questionnaires or written comments (n= 4), think- aloud protocols (n= 1), semi-structured interviews (n= 10), and focus group discussions (n= 4). Observations (n= 5) and research of document sources (n= 7) were conducted to collect additional information. The method of "phenomenography" (Marton, 1981) has been frequently used to analyse the qualitative data gathered. Differences in conceptualisation are systematically explored by a rigorous procedure in which the transcripts are categorised in a relatively small number of recognisable different categories, independently checked by another researcher. This procedure strengthens the value of this qualitative research, and allows connections to be made with quantitative studies (Entwistle, et al 2001). Most studies have a sample of 101 to 200 subjects (n= 11) and from 31 to 100 persons (n= 9). A relatively high number of studies (n= 6) has a sample size of less than 30 students. Three, and five investigations have respectively a sample of 201 to 300 subjects and more than 300 persons. The sample size of the two case studies (n= 13 cases), is unknown.

5. OVERALL SUMMARY AND CONCLUSIONS

Student learning is subject to a dynamic and richly complex array of influences which are both direct and indirect, intentional and unintended (Hounsell, 1997b). In this review, we had the purpose to investigate students' perceptions about assessment in higher education and its influences on student learning and more broadly, the learning- teaching environment. Following questions were of special interest to this review: (1) what are the influences of the (perceived) characteristics of assessment on students' approaches to learning, and vice versa, (2) what are students' perceptions about different alternative assessment formats and methods, and (3) what are the influences of these students' perceptions about assessment on student learning?

In short, this review evidenced that students' perceptions about assessment and its properties, have considerable influences on students' approaches to learning and more in general, to student learning. Also, vice versa, students' approaches to learning influence the ways in which students' perceive assessment.

Furthermore, it was found that students hold strong views about different formats and methods of assessment. Educational research revealed that within conventional assessment practices, students perceive the multiple choice format as more favourable than the constructed response/ essay items. Especially with respect to students' perceptions on the dimensions of perceived difficulty, lower anxiety and complexity, and higher success expectancy, students give preference to this examination format. Curiously, over the past few years, multiple choice type tests have been the target of severe public and professional attack on various grounds. Indeed, the attitude and semantic profile of multiple choice exams emerging from the examinee's perspective is largely at variance with the unfavourable and negative profile of multiple choice exams often emerging from some of the anti- test literature (Zeidner, 1987). However, within the group of students some remarkable differences are found. For example, students with good learning skills and students with low test anxiety rates, both seem to favour the essay type exams, while students with poor learning skills and low test anxiety have more unfavourable feelings towards this assessment mode. It was also found that this essay type of examination goes together with deep(er) approaches to learning than multiple choice formats. Some studies found gender effects, with females being less favourable towards multiple-choice formats than to essay examinations (Birenbaum & Feldman, 1998).

When students discuss alternative assessment, their perceptions about conventional assessment formats, contradict strongly with the students' more favourable perceptions towards alternative methods. Learners, experiencing alternative assessment modes, think positive about new assessment strategies, such as portfolio assessment, self and peer assessment, simulations. From students' point of view, assessment has a positive effect on their learning and is fair when it (Sambell et al., 1997):
- Relates to authentic tasks.
- Represents reasonable demands.
- Encourages students to apply knowledge to realistic contexts.
- Emphasis the need to develop a range of skills.
- Is perceived to have long- term benefits.

Furthermore, different assessment methods seem to assess various skills and competencies. The goal of the assessment has thus a lot to do with the type of assessment and the consequent impact on students' perceptions. It is important to value each assessment method, within the learning environment for which it is intended, and taking its purposes and skills to be assessed into consideration. It is not desirable to apply new or popular assessment modes, without reflecting upon the characteristics, purposes and criteria of the assessment, and without considering the learning- teaching environment of which the assessment type is only one part shaping and modelling student perceptions. Other influences like characteristics of the student (e.g. the students' motivation, anxiety level, approach to learning, intelligence, social skills, and former educational experiences) and properties of the learning-teaching environment (e.g. characteristics of the educator, the teaching method used, the resources available) have to be included.

The literature and research on students' perceptions about assessment is relatively limited. Besides the relational and semi-experimental studies on students' approaches to learning and studying in relation to students' expectations, preferences and attitudes towards assessment that is well known, especially the research on students' perceptions about particular modes of assessment is restricted. Most results are consistent with the overall tendencies and conclusions. However, some inconsistencies and even contradictory results are revealed within this review. Further research can elucidate these results and can provide us with additional information and evidence on particular modes of assessment in order to gain more insight in the process of student learning. These findings can equip us with valuable information in trying to comply with the more "benign" approach and the pressures that it places in trying to maintain a truly (versus only "perceived") and more valid system of assessment. Many of the research findings and possible solutions to assessment problems are good ideas, but they have to be applied with great care and knowledge of assessment in its full complexity. In this regard, it is important to view the first order perspective and the second order approach to the study of human learning and assessment as complementary. Ultimately, it is the interaction between the two perspectives that leads to the understanding of the assessment phenomenon.

This review has tried and hopefully succeeded to provide educators with an important source of inspiration, namely students' perceptions about assessment and its influences on student learning, which can guide them in their reflective search to improve their teaching and assessment practices, and as a consequence, to achieve a higher quality of education.

REFERENCES

Ashworth, P., & Bannister, P. (1997). Guilty in whose eyes?. University students' perceptions of cheating and plagiarism in academic work and assessment. *Studies in Higher Education, 22* (2), 187-203.
Birenbaum, M. (1997). Assessment preferences and their relationship to learning strategies and orientations. *Higher Education, 33,* 71-84.
Birenbaum, M., & Feldman, R. A. (1998). Relationships between learning patterns and attitudes towards two assessment formats. *Educational Research, 40* (1), 90-97.
Birenbaum, M., Tatsuoka, K. K., & Gutvirtz, Y. (1992). Effects of response format on diagnostic assessment of scholastic achievement. *Applied psychological measurement, 16* (4), 353-363.
Boes, W., & Wante, D. (2001). *Portfolio: the story of a student teacher in development./ Portfolio: het verhaal van de student in ontwikkeling* [Unpublished dissertation/ Ongepubliceerde licentiaatverhandeling]. Katholieke Universiteit Leuven, Department of Educational Sciences.
Challis, M. (2001). Portfolios and assessment: meeting the challenge. *Medical Teacher, 23* (5), 437-440.
Chambers, E. (1992). Work- load and the quality of student learning. *Studies in Higher Education, 17* (2), 141-154.
Dochy, F., Segers, M., Gijbels, D., & Van den Bossche, P. (2002). *Studentgericht onderwijs en probleemgestuurd onderwijs. Betekenis, achtergronden en effecten.* Utrecht: Uitgeverij LEMMA.
Drew, S. (2001). Perceptions of what helps learn and develop in education. *Teaching in Higher Education, 6* (3), 309-331.
Edelstein, R. A., Reid, H. M., Usatine, R., & Wilkes, M. S. (2000). A comparative study of measures to evaluate medical students' performances. *Academic Medicine, 75* (8), 825-833.
Entwistle, N., & Entwistle, A. (1997). Revision and experience of understanding. In F. Marton, D. Hounsell, & N. Entwistle (Eds.), *The experience of learning. Implications for teaching and studying in higher education* [second edition] (pp. 146-158). Edinburgh: Scottish Academic Press.
Entwistle, N. J. (1991). Approaches to learning and perceptions of the learning environment. Introduction to the special issue. *Higher Education, 22,* 201-204.
Entwistle, N. J., & Entwistle, A. (1991). Contrasting forms of understanding for degree examinations: the student experience and its implications. *Higher Education, 22,* 205-227.
Entwistle, N. J., & Ramsden, P. (1983). *Understanding student learning.* London: Croom Helm.
Entwistle, N., McCune, V., & Walker, P. (2001). Conceptions, styles, and approaches within higher education: analytical abstractions and everyday experience. In Sternberg & Zhang, *Perspectives on cognitive, learning and thinking styles* (pp. 103-136). NJ: Lawrence Erlbaum Associates.
Entwistle, N., & Tait, H. (1995). Approaches to studying and perceptions of the learning environment across disciplines. *New directions for teaching and learning, 64,* 93-103.
Flint, N. (2000). Culture club. An investigation of organisational culture. Paper presented at the Annual Meeting of the Australian Association for Research in Education, Sydney.
Franklyn-Stokes, A., & Newstead, S. E. (1995). Undergraduate cheating: who does what and why? *Studies in Higher Education, 20* (2), 159-172.

Friedman Ben-David, M., Davis, M. H., Harden, R. M., Howie, P. W., Ker, J., & Pippard, M. J. (2001). AMEE Medical Education Guide No. 24: Portfolios as a method of student assessment. *Medical Teacher, 23* (6), 535-551.

Hounsell, D. (1997a). Contrasting conceptions of essay- writing. In F. Marton, D. Hounsell, & N. Entwistle (Eds.), *The experience of learning. Implications for teaching and studying in higher education* [second edition] (pp. 106-126). Edinburgh: Scottish Academic Press.

Hounsell, D. (1997b). Understanding teaching and teaching for understanding. In F. Marton, D. Hounsell, & N. Entwistle (Eds.), *The experience of learning. Implications for teaching and studying in higher education* [second edition] (pp. 238-258). Edinburgh: Scottish Academic Press.

Kniveton, B. H. (1996). Student perceptions of assessment methods. Assessment and *Evaluation in Higher Education, 21* (3), 229-238.

Lomax, R. G. (1996). On becoming assessment literate: an initial look at preservice teachers' beliefs and practices. *Teacher educator, 31* (4), 292-303.

Marlin, J. W. Jr. (1987). Student perception of End-of-Course-Evaluations. *Journal of Higher Education, 58* (6), 704-716.

Marton, F. (1976). On non- verbatim learning. II. The erosion of a task induced learning algorithm. *Scandinavian Journal of Psychology, 17*, 41-48.

Marton, F. (1981). Phenomenography- describing conceptions of the world around us. *Instructional Science, 10*, 177-200.

Marton, F., & Säljö, R. (1997). Approaches to learning. In F. Marton, D. Hounsell, & N. Entwistle (Eds.), *The experience of learning. Implications for teaching and studying in higher education* [second edition] (pp. 39-59). Edinburgh: Scottish Academic Press.

Meyer, D. K., & Tusin, L. F. (1999). Pre-service teachers' perceptions of portfolios: process versus product. *Journal of Teacher Education, 50* (2), 131-139.

Mires, G. J., Friedman Ben-David, M., Preece, P. E., & Smith, B. (2001). Educational benefits of student self- marking of short- answer questions. *Medical Teacher, 23* (5), 462-466.

Orsmond, P., Merry, S., et al. (1997). A study in self- assessment: tutor and students' perceptions of performance criteria. *Assessment and Evaluation in Higher Education, 22* (4), 357-369.

Ramsden, P. (1981). *A study of the relationship between student learning and its academic context* [Unpublished Ph.D. thesis]. University of Lancaster.

Ramsden, P. (1997). The context of learning in academic departments. In F. Marton, D. Hounsell, & N. Entwistle (Eds.), *The experience of learning. Implications for teaching and studying in higher education* [second edition] (pp. 198-217). Edinburgh: Scottish Academic Press.

Richardson, J. T. E. (1995). Mature students in higher education: II. An investigation of approaches to studying and academic performance. *Studies in Higher Education, 20* (1), 5-17.

Rickinson, B. (1998). The relationship between undergraduate student counseling and successful degree completion. *Studies in Higher Education, 23* (1), 95-102.

Säljö, R. (1975). *Qualitative differences in learning as a function of the learner's conception of a task.* Gothenburg: Acta Universitatis Gothoburgensis.

Sambell, K., & McDowell, L. (1998). The construction of the hidden curriculum: messages and meanings in the assessment of student learning. *Assessment and Evaluation in Higher Education, 23* (4), 391-402.

Sambell, K., McDowell, L., & Brown, S. (1997). 'But is it fair?': an exploratory study of student perceptions of the consequential validity of assessment. *Studies in Educational Evaluation, 23* (4), 349-371.

Sarason, I. G. (1984). Stress, anxiety and cognitive interference : reactions to tests. *Journal of Personality and Social Psychology, 46* (4), 929-938.

Segers, M., & Dochy, F. (2001). New assessment forms in Problem- based Learning: the value- added of the students' perspective. *Studies in Higher Education, 26* (3), 327-343.

Slater, T. F. (1996). Portfolio assessment strategies for grading first- year university physics students in the USA. *Physics Education, 31* (5), 329-333.

Spielberger, C. D., Gorsuch, R. L., & Lushene, R. E. (1970). STAI manual for a state- trait anxiety inventory. California: Consulting Psychologist Press.

Topping, K. (1998). Peer assessment between students in colleges and universities. Review of *Educational Research, 68* (3), 249-276.

Treadwell, I., & Grobler, S. (2001). Students' perceptions on skills training in simulation. *Medical Teacher, 23* (5), 476-482.

Trigwell, K., & Prosser, M. (1991). Improving the quality of student learning: the influence of learning context and student approaches to learning on learning outcomes. *Higher Education, 22*, 251-266.

Traub, R. E., & MacRury, K. (1990). Multiple choice vs. free response in the testing of scholastic achievement. In K. Ingenkamp & R. S. Jager (Eds.), *test und tends 8: jahrbuch der pädagogischen diagnostik* (pp. 128-159). Weinheim und Base: Beltz Verlag.

Van IJsendoorn, M. H. (1997). Meta- analysis in early childhood education: progress and problems. In B. Spodek, A. D. Pellegrini, & N. O. Saracho (Eds.), *Issues in early childhood education. Yearbook in early childhood education* [Volume 7]. New York: Teachers College Press.

Van Rossum, E. J., & Schenk, S. M. (1984). The relationship between learning conception, study strategy and learning outcome. *British Journal of Educational Psychology, 54* (1), 73-83.

Zeidner, M. (1987). Essay versus multiple-choice type classroom exams: the student's perspective. *Journal of Educational Research, 80* (6), 352-358.

Zoller, U., & Ben-Chaim, D. (1988). Interaction between examination-type anxiety state and academic achievement in college science: an action- oriented research. *Journal of Research in Science Teaching, 26* (1), 65-77.

Assessment of Students' Feelings of Autonomy, Competence, and Social Relatedness: A New Approach to Measuring the Quality of the Learning Process through Self- and Peer Assessment

Monique Boekaerts & Alexander Minnaert
Center for the Study of Education and Instruction, Leiden University, The Netherlands

1. INTRODUCTION

Falchikov and her colleagues (Falchikov, 1995; Falchikov & Boud, 1989; Falchikov & Goldfinch, 2000) differentiated between self-assessment and peer assessment and illustrated that student involvement in assessment typically requires them to use their own criteria and standards to make their judgments. Falchikov and Goldfinch maintained that student assessment is a clear manifestation of instruction set up according to the principles of social constructivism. This new form of instruction requires students to learn from and with each other. A marked advantage of this socially situated form of instruction is that it naturally elicits peer assessment and self-assessment through reflection and self-reflection, even in the absence of marking and grading. It is encouraging that both meta-analytic studies on assessment in higher education, namely the one conducted by Falchikov and Boud on self-assessment and the more recent review conducted by Falchikov and Goldfinch on peer assessment, confirmed that students' assessments are more accurate when the criteria for judgement are explicit and well understood. This finding does not come as a surprise since self-assessment and peer assessment are new skills that students must acquire and learn to use in the context of skill acquisition. On comparing the outcomes of the two meta-analyses, Falchikov and Goldfinch came up with an intriguing difference.

They revealed that in the self-assessment study the students' ratings in high level courses were more similar to their teachers' than in the low level courses. Also, student assessors showed more agreement with their teachers in the area of science. Neither course-level differences nor subject area differences were found in the peer-assessment study. Falchikov and Goldfinch were surprised that senior students who are supposed to have a better understanding of the criteria by which they judge performance in a domain and also had more practice in peer assessing did not outperform their juniors. They suggested that the lack of differentiation between beginning and more advanced students is due to the public nature of peer assessment. Assessing one's own performance is usually done in private using one's own internal standards. This may be a more difficult task to do than comparing the public performance of one's peers and ranking their performance or skill acquisition process in ascending or descending order.

We think that this finding offers two lessons to theorists on assessment. First, the results point to the fact that socially situated assessment (i.e., assessment that takes place in the context of the peer group) whether it is assessment of one's own performance or the assessment of a group member's performance, is totally different from self-assessment in relation to individual work. Assessment done in public implies that social expectations and social comparisons contribute significantly to one's judgement. For this reason it is highly important that assessment researchers take care not to lump together these four different forms of assessment (self-assessment and peer assessment of individual work and self-assessment and peer assessment of collaborative work).

The second lesson that assessment researchers should draw from these findings is that motivation factors are powerfully present in any form of assessment and bias the students' judgement of their own or somebody else's performance. In a recent study on the impact of affect on self-assessment, Boekaerts (2002, in press) showed that students' appraisal of the demand capacity ratio of a mathematics task, before starting on the task, contributed a large proportion of the variance explained in their self-assessment at task completion. Interestingly, the students' affect (experienced positive and negative emotions during the math task) mediated this effect. Students who experienced intense negative emotions during the task underrated their performance while students who experienced positive emotions, even in addition to negative emotions, overrated their performance. This finding is in line with much research in mainstream psychology that has demonstrated the effect of positive and negative mood state on performance and decision-making.

On scanning the literature on assessment, we were surprised that most of the reported studies are largely concerned with peer assessment and self-assessment of marking and grading. For example, in the literature on higher education, students who follow courses in widely different subject areas are typically asked to complete different instruments to assess diverse aspects of performance, including interpersonal skills and professional practice (e.g., poster and oral presentation skills, class participation, global peer performance, global traits displayed in group discussion or dyadic interaction, practical tests and examinations, videotaped interviews, tutorial problems, counselling skills, critiquing skills, internship performance, simulation training, ward assignments, group processes, clinical performance, laboratory reports). We did not come across any study that asked students to assess their own or their peers' interest in skill acquisition or professional practice. Nevertheless, we are of the opinion that students' interest in skill acquisition biases their self-assessment and peer assessment, and therefore endangers the validity and reliability of the assessment procedure. In order to investigate this claim, it is important that instruments become available that provide a window on the students' interest in the skill acquisition process.

Why is it important to assess students' interest in relation to skill acquisition within a domain? The results from a wide range of recent studies show that interest has a powerful, positive effect on performance (for a review, see Hoffman, Krapp, Renninger, & Baumert, 1998; Schiefele, 2001). This positive effect has been demonstrated across domains, individuals, and subject-matter areas. Moreover, interest has a profound effect on the quality of the learning process. Hidi (1990) documented that students, who are high on interest do not necessarily spend more time on tasks but the quality of their attentional and retrieval processes, as well as their interaction with the learning material is superior, compared to students low on interest. They use less surface level processing, such as rehearsal, and more deep level processing, such as elaboration and reflection. In other words, interest is a significant factor affecting the quality of performance and should therefore be considered when interpreting students' outcomes, namely their self-assessment and peer assessment. In light on these results, it is indeed surprising that the literature on assessment and on the qualities of new modes of assessment mainly focuses on the assessment of performances and does not take account of the students' assessment of the underlying factors of performance, such as personal interest in skill development and satisfaction of basic psychological needs. Nevertheless, it is clear that instruction situations that present students with learning activities that satisfy their basic psychological needs, create the conditions for interest to develop. Students can give valuable information on the factors underlying their interest, and as such help teachers to create more powerful learning environments.

2. ASSESSING THE STUDENTS' INTEREST IN SKILL ACQUISITION

2.1 Three Basic Psychological Needs: Autonomy, Relatedness, and Competence

Several researchers, amongst others Deci and Ryan (1985), Deci, Vallerand, Pelletier, and Ryan (1991), Ryan and Deci (2000) and Connell and Wellborn (1991) provided evidence that specific factors in social contexts also produce variability in student motivation. They theorised that learning in the classroom is an interpersonal event that is conducive to feelings of social relatedness, competence, and autonomy. On the basis of their extensive research, Deci and Ryan argued that students have three basic psychological needs: they want to feel competent during action, to have a sense of autonomy and to feel that one is secure and has established satisfying relationships. Deci and his colleagues further argued that intrinsic motivation, which is a necessary condition for self-regulation to develop, is facilitated when teachers support rather than thwart their students' psychological needs. More concretely, they predicted that intrinsic motivation develops by providing optimal challenges for one's students, providing effectance, promoting feedback, encouraging positive interactions between peers, and keeping the classroom free from demeaning evaluations.

Deci and Ryan's influential work provided insights into the reasons behind students' task engagement in the classroom. They linked students' satisfaction of basic psychological needs to their engagement patterns, locating regulatory styles along a continuum ranging from amotivation, or students' unwillingness to cooperate, to external regulation, introjection, identification, integration, and ultimately active personal commitment or intrinsic motivation. Evidence to date suggests that students do not progress through each stage of internalisation to become self-regulated learners within a particular domain. Prior experiences and situational factors influence this process. As far as the situational factors are concerned, Ryan and his colleagues (e.g., Ryan, Stiller, and Lynch, 1994) showed that students need to be respected, valued, and cared for in the classroom in order to be willing to accept school-related behavioural regulation (see also Battistich, Solomon, Watson, & Schaps, 1997). They also want to satisfy their need for social relatedness, and their need to feel self-determined. Williams and Deci's (1996) longitudinal study showed that in order to become self-regulated learners, in the true sense of the word, students need teachers who are supportive of their competency development and their sense of

Assessment of the Quality of the Learning Process 229

autonomy. Deci, Egharari, Patrick, and Leone (1994) clearly showed that autonomy support is crucial for self-regulation to develop (i.e., students must grasp the meaning and worth of the learning goal, endorsing it fully and using it to steer and direct their behaviour). It is important to note in this respect that students who work in a controlling context may also show internalisation, provided the social context supports their competency development and social relatedness. However, under these conditions, internalisation is characterised by a focus on approval from self or others (introjection).

2.2 Assessing Feelings of Autonomy, Competence, and Relatedness On-line

Despite this interesting research, instruments that help teachers to gain insights into the interplay between students' developing self-regulation, on the one hand, and their need for competence, autonomy, and social relatedness, on the other, are still rare. Yet, such instruments are essential to help teachers create a learning environment that is conducive to deep learning in successive stages of a course and to the development of self-regulation. We reasoned that it is crucial that students are invited to set their own goals and to direct their learning towards the realisation of these goals, yet perceive that the teacher is supportive of their autonomy, competence development, and social relatedness. This is particularly true for students who are working in cooperative learning environments with the teacher as a coach. By implication, it is important that teachers gain insight into how their students interpret the learning environment. This information will help them to increase or decrease task demands, external regulation (or scaffolding), and social (in)dependence in a flexible way. Ideally, teachers should develop antennae to pick up such signals. In order to help teachers to grow these antennae, we constructed an instrument that registers how individual group members value the learning environment in terms of the autonomy it grants, in terms of the perceived feeling of belonging, and in terms of their competency development. We reasoned that, students who are working on self-chosen group projects for several weeks:
1. are aware of their feelings of autonomy, competence and social relatedness,
2. can report on these feelings, and
3. can use these feelings as a source of information for determining how interested they are in the group project.

We predicted that feelings of competence, autonomy, and relatedness fluctuate during the course of a group project and have a strong impact on reported personal interest during successive stages of the project. We also reasoned that positive and negative perceptions of constraints and affordances at any point in time interact and jointly determine whether students appraise the current learning opportunity as optimal or sub-optimal for group learning. During the course of our work with students in higher education, we had observed many times that undergraduates react differently to external regulation, social support, and scaffolding in the various stages of a learning trajectory. It is easy to imagine that the extent to which students perceive that their fluctuating psychological needs are fulfilled has an impact on their interest in the project, implying that interest also fluctuates over time. Our position is that students who are working in a learning environment that they perceive as "optimal" are willing to invest resources to self-regulate their learning. By contrast, students who perceive the learning context as "sub-optimal" (e.g., not enough structure, no autonomy support) decline the teacher's offer to coach their self-regulation process, mainly because they feel a lack of purpose (no goal-oriented behaviour), low relatedness, and no inclination to engage in learning tasks set by the teacher or by the group. Most teachers refer to this feeling state as: low personal interest in the task or project.

2.3 Constructing the First Version of the Quality of Working in Group Instrument

The focus in this paper is on the construction of the paper-and-pencil version of an instrument that assesses students' feelings of autonomy, competence, and relatedness on-line during successive sessions of working on a group project. In order to test the hypothesis that feelings of autonomy, competence and social relatedness fluctuate during the course of a group project and have a strong impact on personal interest, we needed an instrument that captures the fulfilment of these basic needs on-line. Basically, there are three choices one can make: signal-contingent methods, event-contingent methods, and interval-contingent methods. After careful considerations of the alternatives, it was decided to opt for event-contingent sampling.

The paper and pencil version of the Quality of Working in Group Instrument (QWIGI) was constructed after examining relevant instruments and several try-outs in secondary vocational education and in higher education. QWIGI is a simple instrument that consists of a number of self-report items that can be answered on Likert scales. Completing the questionnaire requires that students stop and think about the quality of the group learning process as they currently perceive it, starting with the particular feature that is highlighted in the item. Based on observations in the college classroom, we predicted that students' sense of autonomy (feeling free to initiate and regulate their own actions) during group work is intricately linked to their understanding of how to solve a problem or complete an assignment and being self-efficacious in performing the necessary and sufficient actions (competence) as well as to their ability to establish satisfying connections with members of the group (relatedness). The relation between student' perception of competence and social relatedness is less clear. A second prediction pertains to personal interest. It was hypothesised that perceived autonomy, competence, and social relatedness jointly influence students' assessment of personal interest. A third prediction concerns the impact that these three predictors have, over time on the assessment of personal interest. In line with our observations in the college classroom, it was predicted that the degree of personal interest that students express in a group assignment could best be explained at the start of the project and just before finishing the project.

3. RESEARCH METHOD

3.1 Subjects

Participants were 54 undergraduate students who participated in a course in Educational Psychology that lasted several weeks and was taught according to the principles of social constructivism. The vast majority of the students were females. Students worked in nine self-selected groups of 5 or 6 students. Data of 4 students with incomplete responses were excluded from all analyses.

3.2 Instrument

The Quality of Working in Groups Instrument consists of a single printed sheet on which 10 bipolar items are presented. Together, these items assess the students' psychological needs: feelings of autonomy (2), social relatedness (2), and competence (2). In addition, their interest in the group project is assessed (2) as well as the degree of responsibility for learning (2). It was decided to include the latter two items because we thought that students' need to develop secure and satisfying relationships with peers is empirically distinct from the degree to which they feel personally responsible for promoting group learning. Each item consists of a five-point-bipolar-Likert-scale with two opposing statements located at either end of the scale. The statements were constructed on the basis of discussions and interviews with similar groups of students in previous years. An example of an item intended to measures students' feelings of autonomy is:

| There is plenty of room for making our own decisions | ooooo ooooo | There is no room for making our own decisions |

The exact wording of the items can be found in the appendix. Students who indicated high agreement with a positive statement (i.e. high feelings of autonomy, competence, social relatedness, and interest, respectively) received a score of 5 whereas high agreement with the negative statements received a score of 1.

3.3 Procedure

The course lasted 12 weeks and was split up into two consecutive units. The first unit involved five three-hour sessions that each started with direct teaching followed by group work. Students had to prepare for class and used this material when working in self-chosen groups of five or six students. They worked on parallel or complementary assignments and performed one of the rotating roles (chairperson, written report secretary, verbal report secretary, resource manager, ordinary member). At the end of each session, the verbal report secretary presented the group solution in public and the teacher invited all students to decontextualise the presented solutions. The first unit was completed with a written exam. The second unit of the course also consisted of five three-hour sessions. Unlike in the first unit, students did not have to take an exam but had to write a group paper that would result in a group mark. They were told that the group paper should focus on a specific self-regulatory skill that they wanted to improve in primary school students. In order to build up the competence of all the group members, students had to read specific articles for each session, visit primary schools and observe relevant classroom sessions, and enter into dialogue with group members in order to construct their own opinion of the merit of the intervention modules they had read about. All groups were free to select the domain (math, reading, writing, etc.) as well as the types of metacognitive or motivation skill(s) they wanted to improve. They had to set their own goals and monitor their progress to the goal. In order to help them structure and organise the group activities and the preparation for the paper, teacher-guided discussions were organised at the beginning of each three-hour session. The literature covered for that session was discussed and the teacher also provided a framework for interpreting this information and for linking it to the topics discussed in previous sessions. For the remaining time of the three-hour sessions, students worked on their own project and the teacher provided feedback on the group activities and on the preliminary table of contents for the paper. The QWIGI was handed to the resource managers before students started on their group project and they were asked to hand it to the group members after about one hour into the group work. The resource manager collected the completed questionnaires and dropped them in a box that was located at the exit of the room. Students were highly compliant with the request to complete the questionnaires.

3.4 Assessing the Reliability and Validity of QWIGI

Falchikov and Goldfinch (2000) were concerned with the reliability and validity of the assessment instruments used in the studies that were included in their review. They explained that high peer-faculty agreement is the best indicator of "validity" of an instrument, whereas high agreement between peer ratings is the best indicator of the "reliability" of an assessment instrument. As previously mentioned, the assessment instruments reviewed in Falchikov and Goldfinch's meta-analysis mainly concern marking and grading. Clearly, marks and grades used in educational settings are neither very reliable nor very valid indicators of achievement, even when there is reasonable agreement between various raters. It is generally accepted that multiple ratings are superior to single ones (Fagot, 1991) because the ratio of true score variance to error variance is increased. Likewise, when group members are asked to judge the performance of each participating student, the reliability of the average scores is increased with the number of raters, but group size should be kept small to avoid the social-loafing effect (Latane, Williams & Hawkins, 1979). Contrary to the assessment instruments reviewed in Falchikov and Goldfinch's study, our instrument does not deal with peer assessment or self-assessment of performance. Rather, it concerns peer assessment and self-assessment of the quality of the learning process within a collaborative learning context and the impact of these perceptions have on personal interest in the project. It is our basic tenet that students' profiles (i.e. their score on perceived autonomy, competence, and relatedness) are of crucial importance with respect to their assessment of personal interest in skill acquisition.

Contrary to what is acceptable practice concerning the validity of performance assessment, faculty ratings are not more accurate than student ratings in relation to the assessment of personal interest. On the contrary, students are better judges of their personal interest than faculty ratings. To evaluate the construct validity of the items at the different measurement points, confirmatory factor analyses with LISREL (Jöreskog & Sörbom, 1993) were performed. The hypothesis tested was that perceived autonomy, competence, and social relatedness jointly impact on students' personal interest and the impact of these predictors over time was estimated with multiple regression analyses.

To evaluate the reliability of the QWIGI scores, neither classical item analysis nor high agreement between peer ratings are considered appropriate estimations of the reliability of the scale. Classical item analysis is considered inappropriate due to the restricted number of items per scale (namely 2). High agreement between peer ratings is regarded as an inappropriate indicator of reliability due to the presupposed process-related fluctuations in students' psychological needs during the course of a group project. As mentioned previously, we wanted to link students' profiles on perceived autonomy, competence, and relatedness to their developing interest in the group project. We therefore decided to calculate profile reliability for all the scales, using Lienert and Raatz's (1994, p. 324) formula. These researchers established that profile reliability is stronger when the reliability of the separate scales is high and the inter-correlations between the scales are low. Lienert and Raatz mentioned a correlation coefficient of .50 as the lower limit of sufficiency.

4. RESULTS

In Table 1 the mean scores and standard deviations for autonomy, competence, social relatedness, and interest are given for the total group and for the nine groups, separately for the five measurement points (one each session). Using the formula provided by Lienert and Raatz, profile reliability was calculated. It was more than sufficient for further use in the context of a course that was taught according to the principles of social constructivism: profrtt=.71.

Assessment of the Quality of the Learning Process

Table 1. Means and standard deviations for autonomy, competence, social relatedness, and interest by group and by measurement moment

Group	Autonomy M	Autonomy SD	Competence M	Competence SD	Social relatedness M	Social relatedness SD	Interest M	Interest SD
Moment 1 (orientation stage)								
Bookies	8.00	1.41	6.60	2.30	8.60	0.96	8.80	1.64
Celsius	7.00	2.45	6.40	1.67	8.20	0.57	7.20	1.48
Esprit	7.00	1.15	6.25	0.50	7.62	1.43	7.75	2.63
Group3	6.50	0.84	7.17	0.98	8.08	0.66	4.67	1.63
Lot	8.33	2.08	6.00	1.00	8.16	0.57	8.67	1.53
The crazy chicks	8.00	1.00	6.67	1.15	8.16	1.15	10.00	0.00
Manpower	7.33	2.42	7.00	1.63	8.07	1.90	8.57	1.40
The knife fighters	7.00	2.10	6.67	1.21	8.00	1.09	8.83	0.75
Skillis	8.20	1.30	4.60	1.95	9.40	1.08	10.00	0.00
Total	7.38	1.72	6.43	1.56	8.26	1.16	8.14	2.08
Moment 2 (execution stage)								
Bookies	8.40	2.19	5.60	1.67	8.70	0.75	8.80	0.84
Celsius	7.20	2.68	7.20	2.28	7.20	0.44	8.00	1.41
Esprit	8.33	0.82	6.67	1.63	7.91	1.46	8.67	1.51
Group3	8.17	1.17	7.33	1.03	8.50	0.63	7.83	1.47
Lot	9.75	0.50	5.50	1.91	8.12	1.10	6.25	2.63
The crazy chicks	6.33	1.53	6.33	1.53	7.50	1.32	7.00	2.65
Manpower	6.00	2.16	7.86	1.46	7.64	1.90	8.14	1.95
The knife fighters	6.20	1.30	6.80	2.17	8.30	1.25	8.80	0.45
Skillis	7.80	1.30	6.40	1.14	9.30	0.75	10.00	0.00
Total	7.54	1.92	6.74	1.69	8.14	1.25	8.26	1.72
Moment 3 (execution stage)								
Bookies	7.67	2.07	7.50	1.52	8.83	1.21	8.83	1.60
Celsius	7.00	2.28	7.17	2.23	9.16	0.81	8.67	1.21
Esprit	8.67	1.51	7.83	1.47	8.75	1.03	9.67	0.52
Group3	7.50	1.76	5.33	2.73	7.60	1.19	4.00	2.10
Lot	7.75	0.50	6.75	2.06	8.25	0.86	8.50	1.29
The crazy chicks	7.40	0.89	7.00	1.83	8.40	0.74	8.60	1.14
Manpower	8.29	1.98	7.86	1.07	7.85	1.43	8.29	1.11
The knife fighters	6.75	2.36	6.50	0.58	7.50	1.68	9.25	0.96
Skillis	8.60	1.95	7.00	2.24	8.70	1.56	9.40	0.89
Total	7.78	1.78	7.04	1.87	8.37	1.13	8.29	2.05

	Moment 4 (execution stage)							
Bookies	7.67	2.07	7.50	1.52	8.83	1.21	8.83	1.60
Celsius	6.50	2.35	6.60	1.52	7.66	1.12	7.67	1.86
Esprit	8.67	2.16	6.67	1.63	8.60	1.38	8.83	1.60
Group3	7.17	2.14	8.67	1.97	7.75	0.98	7.00	1.90
Lot	8.33	2.07	7.00	1.79	8.08	1.62	9.67	0.52
The crazy chicks	8.40	1.52	7.60	1.14	8.30	0.44	8.40	1.34
Manpower	6.17	1.60	7.67	1.03	6.25	1.57	7.33	1.97
The knife fighters	6.20	1.48	5.80	1.48	8.40	0.74	7.60	1.82
Skillis	9.25	1.50	6.25	1.71	8.16	0.28	8.25	2.87
Total	7.54	2.06	7.14	1.65	7.96	1.31	8.18	1.81
	Moment 5 (wrapping up stage)							
Bookies	8.00	1.58	8.20	1.30	9.40	0.65	9.40	0.89
Celsius	7.00	2.00	6.67	1.97	6.08	2.81	7.17	2.79
Esprit	9.00	1.00	7.60	0.55	8.60	1.38	10.00	0.00
Group3	-	-	-	-	-	-	-	-
Lot	-	-	-	-	-	-	-	-
The crazy chicks	8.33	1.21	8.00	1.10	8.41	0.73	8.17	2.14
Manpower	7.50	2.07	8.50	1.38	5.75	1.94	8.50	1.22
The knife fighters	-	-	-	-	-	-	-	-
Skillis	9.25	1.50	7.00	0.82	9.37	0.47	10.00	0.00
Total	8.09	1.69	7.69	1.38	7.78	2.14	8.75	1.85

The internal structure of the questionnaire was examined by confirmatory factor analysis on the competence, autonomy and relatedness items, separately for the five measurement points. We presumed that the items had only substantial loadings on the intended factors. The matrix with intercorrelations between the latent factors was set free because the constructs are not presumed to act independently of each other. The confirmatory factor analyses with Maximum Likelihood estimations yielded good indices of fit, with overall (GFI), incremental (IFI), and comparative (CFI) goodness-of-fit measures ranging between .91 and .99. The $c2/d$ ratios varied between 0.74 and 2.05, which according to Byrne (1989) is evidence of a good fit between the observed data and the model. All items had significant, high loadings on the intended factors, stressing the internal validity of the items involved. With respect to the construct validity of

Assessment of the Quality of the Learning Process

QWIGI, it is concluded that although a very restricted number of items was used to measure competence, autonomy, social relatedness (including the degree of responsibility for learning), and interest, the intended factors were unequivocally retrieved.

The correlations between the three basic psychological needs (i.e., competence, autonomy, and social relatedness), and between these needs and personal interest are printed in Table 2, separately for each measurement point.

Table 2. Pearson correlations between autonomy, competence, social relatedness, and interest by measurement moment (N =32 à 50)

		Autonomy	Competence	Social relatedness
Moment 1 (orientation stage)	Competence	.39**		
	Social relatedness	.46**	.15	
	Interest	.57**	-.03	.33**
Moment 2 (execution stage)	Competence	.21		
	Social relatedness	.38**	.12	
	Interest	.16	.39**	.41**
Moment 3 (execution stage)	Competence	.25*		
	Social relatedness	.13	.32*	
	Interest	.27*	.38**	.39**
Moment 4 (execution stage)	Competence	.06		
	Social relatedness	.41**	.18	
	Interest	.41**	.10	.41**
Moment 5 (wrapping up stage)	Competence	.41**		
	Social relatedness	.57**	.22	
	Interest	.61**	.37**	.56**

* p ≤ .05; ** p ≤ .01

Close inspection of the patterns of correlations revealed that there are basically three stages in the project, namely an orientation stage that is quite short (1 session), a wrapping-up stage that involves the last session and probably, for some groups, the penultimate session. The intermediate stage spans two to three weeks. Examination of these correlational data reveals that, as predicted, autonomy is associated with both competence and social relatedness, as well as with interest in the orientation and wrapping up stage. The correlations between competence and social relatedness are low to modest in all stages of the project, except on measurement point 3 (.32). It is noteworthy that in the last session of the execution stage, autonomy and competence are not associated. This implies that students, who scored high or low on perceived competence, expressed a sense of low autonomy, and vice versa. At the same measurement point, we note that autonomy has a moderate association (.41) with social relatedness, meaning that students high or low on social relatedness express a sense of low autonomy.

A series of multiple regression analyses were conducted to examine how much variance could be explained in personal interest by the three predictors set at the various measurement points.

Table 3. Common and unique effects of autonomy (A), competence (C), and social relatedness (SR) on interest per measurement moment

	Predictor(s)	Zero-order correlation	Semi-partial correlation	R
Moment 1	A	.57**	.54**	
(orientation stage)	C	-.026	-.27*	
	SR	.33**	.06	
	A+C+SR			.64**
Moment 2	A	.16	-.06	
(executive stage)	C	.39**	.35**	
	SR	.41**	.37**	
	A+C+SR			.54**
Moment 3	A	.27*	.17	
(executive stage)	C	.38**	.22	
	SR	.39**	.27*	
	A+C+SR			.50**
Moment 4	A	.41**	.27*	
(executive stage)	C	.10	.03	
	SR	.41**	.26*	
	A+C+SR			.49**
Moment 5	A	.61**	.27*	
(wrapping up stage)	C	.37**	.14	
	SR	.56**	.26*	
	A+C+SR			.67**

* $p \leq .05$; ** $p \leq .01$

Table 3 shows the amount of variance explained in interest, the multiple correlations between the joint psychological needs and interest, and the unique effects of the psychological needs on interest. Most variance was explained in the wrapping-up stage (45%), followed by the orientation stage (40%). The amount of variance explained in the execution stage decreased over time. Interestingly, students' feeling of competence did not contribute unique variance to personal interest, except in the orientation stage

Remarkably, being self-efficacious when starting on the group project affected personal interest negatively, but having a sense of autonomy contributed a large portion of unique variance to personal interest. Please note that social relatedness did not contribute unique variance to interest in the project in the orientation stage. This is easy to understand because the students did not know yet whether the other group members would feel committed to the project, resulting thus in satisfying relationships. In all successive stages, the semi-partial correlations for social relatedness reached significance. Autonomy is predictive of interest on the last three measurement points of working on the project, but not on measurement point 2.

To examine whether the self-selected groups differed significantly on autonomy, competence, social relatedness, and interest, a MANOVA was run. This analysis was followed by a Games-Howell post hoc test to examine which groups differed significantly on the four dependent measures (see Table 4). This type of post hoc test was preferred because homogeneity of variances on the three psychological needs and on interest were not assumed in the different groups. Based on the multivariate tests, it is noteworthy that the groups differed most in the orientation stage. In this stage the groups differ mainly in interest: the post hoc tests indicate that the group scores on interest of five groups (i.e., "bookies", "the crazy chicks", "manpower", "the knife fighters", and "skillis") exceed significantly (">") one specific group (i.e., "group3"). During the execution stage, the multivariate significance drops from moment 2 to moment 4. In this stage, autonomy, social relatedness, and interest play a role at distinct moments during project work. In the wrapping-up stage, multivariate significance increased, indicating that group differences increased. The group effect on the psychological need, social relatedness, at the last two measurement points is remarkable.

Table 4. Manova of group effect on autonomy (A), competence (C), social relatedness (SR), and interest (I) per measurement moment

	Multivariate tests Pillai's Trace	F	Between-subjects group effect A F	A R^2	C F	C R^2	SR F	SR R^2	I F	I R^2	Games-Howell post hoc tests ($a=.05$)	
Moment 1	1.31	2.00**	0.60	.13	1.55	.27	1.18	.22	8.14**	.66	For I:	Bookies, the crazy chicks, manpower, the knife fighters, skillis > group 3
Moment 2	1.24	2.08**	2.65*	.36	1.15	.20	1.46	.24	2.10	.31	For A:	Lot > manpower, the knife fighters
Moment 3	1.01	1.60*	0.80	.14	0.69	.13	1.21	.20	7.83**	.62	For I:	Esprit, skillis > group 3
Moment 4	1.00	1.58*	1.33	.22	1.63	.26	2.29*	.32	1.31	.22	For SR:	No two groups differed significantly
Moment 5	1.12	2.03**	1.43	.22	1.63	.24	5.18**	.50	2.41	.32	For SR:	Bookies, skillis > manpower

* $p \leq .05$; ** $p \leq .01$

5. DISCUSSION

In the introduction, we remarked that motivation factors, including affect and interest, are powerfully present in any learning situation. Three points were made. First, interest is a significant factor affecting the quality of performance and should therefore be considered when interpreting students' outcomes and their self-assessment. Second, affect experienced during the learning situation impacts on self-assessment. Third, new forms of instruction may or may not provide the conditions that satisfy students' basic psychological needs. Students can give valuable information on the factors underlying their interest in a domain or activity (satisfaction of their psychological needs) and this information can help the teacher coach the learning process.

As argued previously, attempts to study students' feelings of autonomy, competence, and social relatedness in close connection with their developing interest are rare. We reasoned that university students who are working on self-chosen projects for several weeks are aware of their feelings of autonomy, competence and social relatedness and can report this information. We also hypothesised that college students use this information when assessing their personal interest in a group project. The focus in this paper was on the construction of an instrument that assesses university students' interest on-line during successive sessions of working on a group project. We predicted and found that feelings of competence, autonomy, and social relatedness fluctuate during the course of the group project and that satisfaction of these basic psychological needs has a strong impact on the personal interest students express in the project during the successive stages. Furthermore, our results suggest that expressed personal interest in a group project is, to a large extent, determined by the student's need satisfaction. In other words, when students express low interest in a group learning project it is advisable that teachers take a closer look at the reasons why their psychological needs are not satisfied because their needs act as signposts on the way to the students' developing personal interest. In Figures 1a and 1b we have visualised the relation between personal interest and the three underlying psychological need states for two groups, namely Celsius and Skillis. As can be seen in these figures, the curve depicting social relatedness is closely linked to expressed personal interest in both groups. In the Celsius group, the curves for need of autonomy and competence are intertwined and influence interest jointly. In the Skillis group, autonomy seems to affect interest separately from students' need of competence.

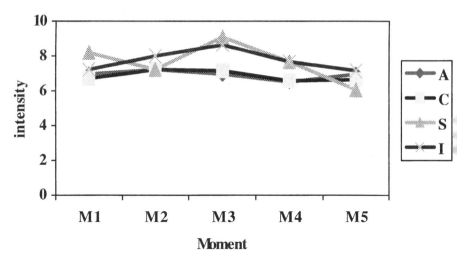

(a) Group Celsius

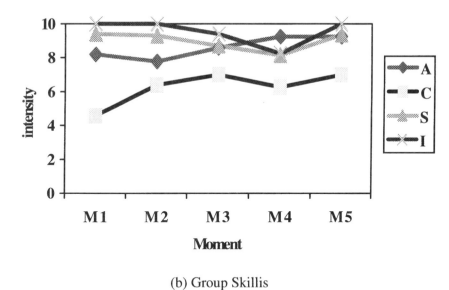

(b) Group Skillis

Figure 1. Relation of interest scores (I) to the three psychological need states (Autonomy, Competence, and Social relatedness) over time for group Celsius (a) and group Skillis (b)

In line with Krapp's (2002) reasoning, we define personal interest in a group project as a relational construct that describes the person-object relation as it is represented in the students' mind. We suggest that students' interest in a group project is a non-linear process that takes place in multiple overlapping contexts. At least two contexts can be discerned. The first context is idiosyncratic in the sense that it is based on the student's motivational beliefs about the domain. These beliefs, which are the result of previous encounters with similar content, are fed into the current situation by activating long-term memory schemata. This idiosyncratic context should be differentiated from a more socially determined context that represents students' current perceptions of group processes and relations.

Which lessons can assessment researchers draw from this study? What does the study mean for ongoing research into new modes of assessment? We found that the students' satisfaction of their psychological needs explained most variance in reported interest at the beginning and end of the project. It seems that students' assessment of the conditions for learning in terms of their perception of autonomy, need for competence, and social relatedness is a good predictor of their interest. We do not want to run the risk of over-interpreting the data from a single study with a limited sample and many potential biases. Hence, we will not compare in detail the pattern of the psychological needs across the various data collection points. Suffice it to draw the reader's attention to the semi-partial correlations recorded in the orientation and the wrapping up stage that are reported in Table 3. As can be seen from this table, satisfaction of the need of autonomy is very important for developing personal interest in the initial stages of the project. The need to satisfy social relatedness does not seem to contribute (in terms of unique variance) much to personal interest expressed in the project and the need to satisfy competence has an inverse relationship with interest at this stage. At the end of the project, the need to satisfy autonomy is less important than in the beginning. Social relatedness now contributes much more to personal interest expressed in the project and the need to satisfy competence has now a modest, unique contribution to interest. This finding suggests that the pattern of the factors that underlie student interest in the project fluctuate during the project and influence each other. It is evident that beginners in a specific domain differ from experts in the psychological needs that they want to satisfy most urgently in order to express interest in an activity. It is highly likely that the pattern between the three basic psychological needs changes when students have discovered for themselves that learning with and from each other does increase their competence.

Our position is that students assess the learning conditions when they are confronted with a new assignment and continue to assess these conditions during their actual performance on the assignment in terms of the satisfaction of their basic psychological needs. This assessment affects the way they reflect on their performance and their judgement of progress (self-assessment). In other words, the students' perception of the learning conditions in terms of the satisfaction of their psychological needs is important for interest assessment (and interest development) as well as for skill assessment (and skill development).

Our view is in accordance with Kulieke et al.'s (1990) proposal to redefine the assessment construct. These researchers suggested that several assessment dimensions should be considered, amongst others, the registration of the extent to which the dynamic learning process was assessed on-line. Researchers and teachers who are involved in collaborative research applauded this new assessment culture, for they are in need of tools that inform the students whether their investment in a course or assignment results in deeper understanding of the content. Our argument is that it is not only skill development that should be assessed (self-assessed, peer-assessed, and teacher assessed) but also the students' developing interest in skill development. Indeed, we believe that students will (continue to) invest resources in skill development, provided they realise that their personal investment leads to valued benefits, such as intrinsic motivation, career perspectives, and personal ownership of work. Ultimately, what gets measured gets managed.

In the study reported here, the information that became available through the self-report data was not fed back to the students during the project. On the basis of the satisfactory results reported here we decided to transform the paper and pencil version of the QWIGI into a computer-based instrument. A digital version of the questionnaire allows us to visualise the waxing and waning of a student's basic psychological needs as well as his or her assessment of developing interest in the group project. Allowing students to inspect the respective curves that depict various aspects of their self-assessment and inviting them to reflect on the reasons behind their self-assessment is a powerful way to confront them with their perception of the constraints and affordances of the learning environment. The computerised version of the questionnaire also allows students to inspect each group member's curves and gain information on how their peers perceive the quality of the learning environment and how interested they are in the group project. We think that the QWIGI is particularly suited for students who are not yet familiar with group projects and for students who express low personal interest in learning from and with each other. The digital version of the instrument is currently used in vocational schools. Students enjoy having the opportunity to assess their personal interest and aspects of the learning environment, especially when these assessments are visualised on the screen in bright colours. An additional benefit of the digital version of the instrument is that the detailed information is also available to the students' teachers. Information about students' developing interest in a group project allows teachers to encourage groups of students to focus on those aspects of the learning episodes that are still problematic for them at that point in time. It also allows teachers to change the task demands, provide appropriate scaffolding, or change the group composition when appropriate.

REFERENCES

Battistich, V., Solomon, D., Watson, M., & Schaps, E. (1997). Caring school communities. *Educational Psychologist, 32* (3), 137-151.

Boekaerts, M. (2002). *Students appraisals and emotions within the classroom context.* Paper presented at the annual conference of the American Educational Research Association, New Orleans, April 2002.

Boekaerts, M. (in press). Toward a Model that integrates Affect and Learning, Monograph published by *The British Journal of Educational Psychology.*

Byrne, B. (1989). Multigroup comparisons and the assumptions of equivalent construct validity across groups: Methodological and substantive issues. *Multivariate Behavioural Research, 24*, 503-523.

Connell, J. P., & Wellborn, J. G. (1991). Competence, autonomy, and relatedness: A motivational analysis of self-system processes. In M. R. Gunnar & L.A. Sroufe (Eds.), *Self processes and development. The Minnesota symposia on child psychology, Vol. 23* (pp. 43-77). Hillsdale, NJ: Lawrence Erlbaum.

Deci, E. L., & Ryan, R. M. (1985). *Intrinsic motivation and self-determination in human behavior.* New York: Plenum.

Deci, E. L., Egharari, H., Patrick, B. C., & Leone, D. R. (1994). Facilitating internalisation – The self-determination theory perspective. *Journal of Personality, 62* (1), 119-142.

Deci, E. L., Vallerand, R. J., Pelletier, L. G., & Ryan, R. M. (1991). Motivation and education: The self-determination perspective. *Educational Psychologist, 26*, 325-346.

Fagot, R. F. (1991). Reliability of ratings for multiple judges – intraclass correlation and metric scales. *Applied Psychological Measurement, 15* (1), 1-11.

Falchikov, N. (1995). Peer feedback marking - Developing peer assessment. *Innovations in Education and Training International, 32*, 175-187.

Falchikov, N., & Boud, D. (1989). Student self-assessment in higher education: A meta-analysis. *Review of Educational Research, 59* (4), 395-430.

Falchikov, N., & Goldfinch, J. (2000). Student peer assessment in Higher Education: A meta-analysis comparing peer and teacher marks. *Review of Educational Research, 70* (3), 287-322.

Hidi, S. (1990). Interest and its contributions as a mental resource for learning. *Review of Educational Research, 60*, 549-571.

Hoffman, L., Krapp, A., Renninger, K. A., Baumert, J. (1998). Interest and learning. Proceedings of the Seeon-conference on interest and gender. Kiel, Germany: IPN.

Jöreskog, K. G., & Sörbom, D. (1993). *LISREL 8: User's reference guide.* Chicago, IL: Scientific Software International.

Krapp, A. (2002). An educational-psychological theory of interest and its relation to self-determination theory. In E. L. Deci & R. M. Ryan (Eds.), *The handbook of self-determination research* (pp. 405-427). Rochester: University of Rochester Press.

Kulieke, M., Bakker, J., Collins, C., Fennimore, T., Fine, C., Herman, J., Jones, B. F., Raack, L., & Tinzmann, M. B. (1990). *Why Should Assessment Be Based on a Vision of Learning?* Oak Brook: NCREL.

Latane, B., Williams, K., & Hawkins, S. (1979). Many hands make light the work: The causes and consequences of social loafing. *Journal of Personality and Social Psychology, 37*, 822-832.

Lienert, G. A., & Raatz, U. (1994). *Testaufbau und Testanalyse*. Weinheim/München, Germany: Beltz, Psychologie Verlags Union.

Ryan, R. M., & Deci, E. L. (2000). Self-determination theory and the facilitation of intrinsic motivation, social development, and well-being. *American Psychologist, 55* (1), 68-78.

Ryan, R. M., Stiller, J., & Lynch, J. H. (1994). Representations of relationships to teachers, parents, and friends as predictors of academic motivation and self-esteem. *Journal of Early Adolescence, 14*, 226-249.

Schiefele, U. (2001). The role of interest in motivation and learning. In J. M. Collis & S. Messick (Eds.), *Intelligence and personality: Bridging the gap in theory and measurement* (pp. 163-194). Mahwah, NJ: Erlbaum.

Williams, G. C., & Deci, E. L. (1996). Internalization of biopsychosocial values by medical students: A test of self-determination theory. *Journal of Personality and Social Psychology, 70* (4), 767-779.

Setting Standards in the Assessment of Complex Performances: The Optimized Extended-Response Standard Setting Method

Alicia S. Cascallar[1] & Eduardo C. Cascallar[2]
[1] *Assessment Group International, UK*, [2] *American Institutes for Research, USA*

1. INTRODUCTION

The historical evidence available on the long tradition of assessment programs and methods, takes us back to Biblical accounts and to the early Chinese civil service exams of approximately 200 B.C. (Cizek, 2001). Since then, socio-political and educational theories and beliefs have had determining impact on the definitions used to implement assessment programs and the use of the information derived from them. The determination of standards is a central aspect of this process. Over time, these standards have been defined in numerous ways, including the setting of arbitrary numbers for passing, the unquestioned establishment of criteria by a ruling board, the performance of individuals in relation to a reference group (not always the "same" group or a "fair" group to compare with), and many other criteria. More recently, changes in the understanding of social and educational phenomena have inspired a movement to make assessments more relevant and better adjusted to the educational goals and the personal advancement of those being tested. Simultaneously, the exponential increase of information and the demanding higher levels of complexity involved in contemporary life, require the determination of complex levels of performance for many current assessment needs. The emergence of new techniques and modalities of assessment has made it possible to address some of these issues. These new methods have also introduced new challenges to maintain the necessary rigor in the assessment process. Standard setting methodologies have provided a means to improve the implicit and explicit categorical decisions in testing methods, making these decisions more open, fair, informed, valid and defensible (Mehrens & Cizek, 2001). There is also now the realisation that standards and the cut scores derived from them are not "found", they are "constructed" (Jaeger, 1989). Standard setting methods are used to construct defensible cut scores.

2. STANDARD SETTING METHODS: HISTORICAL OVERVIEW

Standard setting methods have been used extensively since the early 1970's as a response to the increased use of criterion-referenced and basic skills testing to establish desirable levels of proficiency. The Standards for Educational and Psychological Testing (AERA, APA, NCME, 1999) establish that cut scores based on direct judgement should be designed so that these experts bring their knowledge and experience in determining such cut scores (Standard 4.21). During the standard setting process the judgements of these experts is carefully determined so that the consistency of the process and subsequent establishment of standards is appropriate. Shepard (1980) admonishes that standard-setting procedures, particularly for certification purposes, should balance judgement and passing rates:

> At a minimum, standard-setting procedures should include a balancing of absolute judgements and direct attention to passing rates. All of the embarrassments of faulty standards that have ever been cited are attributable to ignoring one or the other of these two sources of information. (p. 463)

Early references to what came to be known as criterion-referenced measurement can be found in John Flanagan's chapter "Units, Scores, and Norms" (Educational Measurement, 1951). He distinguishes between information regarding test content and information regarding ranks in a specific group, both derived from test score information.

Thus he clearly associated content-based score interpretations with the setting of achievement standards. There were no suggestions on how to set these standards, and even Ebel in 1965 (Ebel, 1965; 1972) gives no concrete advice on the setting of passing scores, and discourages doing so.

By the time the second edition of Educational Measurement was published in 1971, standard-setting methodologies were being proposed for the wave of criterion-referenced measures of the time. The term "criterion-referenced" can be traced to Glaser & Klaus (1962) and Glaser (1963) although the underlying concepts (i.e., standards vs. norms, focus on text content) had already been articulated in the literature (Flanagan, 1951). The criterion-referenced testing practice that ensued proved to be a strong push in the growth of standard-setting methodologies.

The most widely known and used multiple-choice standard setting method: the "Angoff method," was initially described in a mere footnote to Angoff's chapter "Scales, Norms and Equivalent Scores" in the second edition of Educational Measurement (Angoff, 1971). The footnote explained the "Angoff Method" as a "systematic procedure for deciding on the minimum raw scores for passing and honours."

Angoff (1971) very concisely described a method for setting standards.

> ... keeping the hypothetical "minimally acceptable person" in mind, one could go through the test item by item and decide whether such a person could answer correctly each item under consideration. If a score of one is given for each item answered correctly by the hypothetical person and a score of zero is given for each item answered incorrectly by that person, the sum of the item scores will equal the raw score earned by the "minimally acceptable person." (p. 514)

To allow probabilities rather than only binary estimates of success or failure on each item, Angoff (1971) explained:

> A slight variation of this procedure is to ask each judge to state the probability that the "minimally acceptable person" would answer each item correctly. In effect, the judges would think of a number of minimally acceptable persons, instead of only one such person, and would estimate the proportion of minimally acceptable persons who would answer each item correctly. The sum of these probabilities, or proportions, would then represent the minimally acceptable score. (p. 515)

Since then, variations on the Angoff Method have been used widely, but most of the current standard setting methods establishing these procedures have dealt mainly with multiple-choice tests (Angoff, 1971; Ebel, 1972; Hambleton & Novick, 1972; Millman, 1973; Plake, Melican & Mills (1991); Plake & Impara, 1996; Plake, 1998; Plake, Impara & Irwin, 2000; Sireci & Biskin, 1992; Zieky, 2001).

In Zieky's Historical Perspective on Standard Setting (Zieky, 2001), he identifies the recent challenges of standard setting as an attempt to address the additional complications of applying standards to constructed-response tests, performance tests and computerised adaptive tests.

3. EXTENDED-RESPONSE STANDARD SETTING METHODS

Many of the new modes of assessment, including so-called "authentic assessments", address complex behaviours and performances that go beyond the usual multiple-choice tests. This is not to say that objective testing methods cannot be used for the assessment of these complex abilities and skills, but constructed response methods many times present a practical alternative. Setting of defensible, valid standards becomes even more relevant for the family of constructed response assessments, which include extended-response instruments.

Several methods to carry out standard settings on extended-response examinations have been used. Faggen (1994) and Zieky (2001) describe the following methods for constructed-response tests:
1. the Benchmark Method,
2. the Item-Level Pass/Fail Method,
3. the Item-Level Passing Score Method,
4. the Test-Level Pass/Fail Method,
5. the Cluster Analysis Method, and
6. the Generalised Examinee-Centred Method.

3.1 Benchmark Method

In this method judges study "benchmark papers" and scoring guides that serve to illustrate the performance expected at relevant levels of the score scale. Once this has been done, judges select papers at the lowest level that they consider acceptable. The judgements are shared and discussed among judges, and they repeat the process until relative convergence is met. Obtained scores are averaged, and the score for the minimum acceptable paper is determined as the recommended passing score.

3.2 Item-Level Pass/Fail Method

In this method judges read each paper and classify them as "passing" or "failing" without having been exposed to the original grades. Then, they discuss the results obtained and collate the papers. Again, this is an iterative process in which judges can revise their ratings. The process yields estimates of the probability that papers at the various score levels are considered "passing" or "failing". The recommended standard is the point at which the probability of classification to each group is .5 (assuming consequences of either misclassification are equal).

3.3 Item-Level Passing Score Method

In this method judges estimate the average score that would be obtained by a group of minimally competent examinees. They accomplish this after considering the scoring rules and scheme, and the descriptions of performance at each score level. The recommended standard is the average estimated score across the various judges.

3.4 Test-Level Pass/Fail Method

In the Test-Level Pass/Fail Method judgements are made based on the complete set of examinee responses to all the constructed-response questions. Of course, for a one-item test, as Zieky (2001) points out, this method is equivalent to the Item-Level Pass/Fail Method previously described. In addition, Faggen (1994) mentions a variant of this method that incorporates the procedure of making ratings of an item dependent on the judgement for the previous response considered.

3.5 Cluster Analysis Method

The cluster analysis of test scores approach (Sireci, Robin, & Patelis, 1999), although useful in identifying examinees with similar scores or profiles of scores, still leaves several problems unsolved. One is the issue of identifying the clusters that belong to the proficiency groups demanded by the standard setting framework of the test. Another unsolved question is the choice of method to apply in using the clusters to set the cutscores.

3.6 Generalised Examinee-Centred Method

In the generalised examinee-centred method (Cohen, Kane, & Crooks, 1999) all of the scores in an exam are used to set the cutscores, with members of the standard-setting panel rating each performance on a scale linked to the standards that need to be set. The method, as described by Cohen et al. (1999) requires the participants "establish a functional relation... between the rating scale and the test score scale" (p.347). Then, the points identified on the rating scale that provide the definition of the category borders, are converted onto the score scale. This process generates cutscores for each of the category borders. Although this method has some advantages, such as the use of all the scores, and the use of an integrated analysis to generate all the cutscores, it becomes questionable when the correlation between the ratings on the scale and the scores of the test is low (Zieky, 2001).

As Jaeger (1994) points out, even these extensions to constructed-response standard-setting methodologies share the theoretical assumption that the tests to which they are applied are unidimensional in nature, and that the items of each test contribute to a summative scale (p. 3). Of course, we know that many of the instruments that are used in performance assessment, are of a very complex nature, and posses multidimensional structures that cannot be captured by a single score of examinee performance, derived in the traditional ways.

4. THE OPTIMISED EXTENDED-RESPONSE STANDARD SETTING METHOD

In order to also deal with multidimensional scales that can be found in extended response examinations the Optimised Extended-Response Standard Setting method (OER) was developed (Schmitt, 1999). The OER standard setting method uses well defined rating scales to determine the different scoring points where judges will estimate minimum passing points for each scale. For example, if an extended response item is to be assigned a maximum of 6 points, each judge is asked to evaluate how many examinees out of 100 (at each level of competence) would obtain a one, a two, a three, a four, a five, and a six (where the total number has to add to 100). Their ratings are then weighed by the rating scale and averaged across all possible points. This average across judges gives the minimum passing score for each level of competence, for the specific item. The average across all items gives the minimum passing score for each level of competence for the total test. In this way, even rating scales that are different by item can be used. In addition, multiple possible passing points or grades can also be estimated. This method thus provides flexibility based on well-defined rating scales. Once the rating scale is well defined, the judges are trained to evaluate minimum standards based on the corresponding rating scale. This insures consistency between the way standards are set and the way the scoring rubric is assigned.

As with the Angoff Method, the OER standard setting method uses judgement of minimum proficiency. Because of this, the training of the judges and the standard setting process needs to be carefully conducted. We propose the following procedures to meet minimum standards in setting cut scores with the OER standard setting method.

4.1 The OER Standard Setting Method: Steps

4.1.1 Selection of Judges

The panel of judges selected should be large and provide the necessary diverse representation across variables such as: geography, culture, specific technical background, etc. Hambleton (2001) mentions that often 15 to 20 panellists are used in typical USA state assessments. Several studies have addressed the relationship between number of raters and the reliability of the judgements (Maurer, Alexander, Callahan, Bailey, & Dambrot, 1991; Norcini, Shea, & Grosso, 1991; Hurtz and Hertz, 1999). Generalizability analyses have shown that usually a number of judges between 10 and 15 produce phi-coefficients in the range of .80 and above. In most practical situations a minimum of 12 judges has been found necessary to provide reliable outcomes in setting standards (Schmitt, 1999).

These judges should be selected from experienced professionals that are cognisant of the population they will make estimates about. For example, in a National Nursing Program, nurses who have between 5-10 years experience and are currently teaching students at the education level and in the particular content area to be tested, would be good candidates as judges for the standard setting session. If the test is to be administered nationally, the representation of the judges needs to also be national. Regional and/or personal idiosyncrasies in terms of content or standards should be avoided (this needs to be continually monitored during the standard setting process). Judges gender and ethnic representation should mimic, as much as possible, the profession or content area being tested. Invitations to potential judges should include a brief description of what is a standard setting, but should reassure them that the process will be carefully explained when they meet. This assures them that all information will be covered at the same time for all participants.

4.1.2 Standard Setting Meeting

All judges should meet together in one room. Under the current state-of-the-art conditions, having all judgements made at the same time, in the same setting, under standard conditions, and with opportunity to interact in a controlled environment, following exactly the same process, insures a minimum degree of standardisation. Although Fitzpatrick (1989) alerts to the potential problems with group dynamics, Kane (2001) points out that the substantial benefits of having the panellists consider their judgements together as a group far outweigh this risk.

Recently, several programs have instituted innovative processes where judges set standards though web-based "meetings" or other technology-based "meetings". These web-based approaches to carry out standard-settings represent the case of the use of a new technology in the implementation of standard setting sessions. Although innovative, and possibly less short term costly, these distributed "cyber meetings" can produce results that are less reliable and consistent, putting more in question a process that is already judgmental in nature. Harvey and Way (1999), describe one such approach. It should be expected that many such applications will appear in the future, but the underlying issues regarding the method to be used in order to determine the necessary borders of the judged categories will remain the same, varying only in the nature of the implementation media. There is no doubt that in the future new advances might make it possible to carry out standard setting sessions at a distance, without loss of quality in the results.

4.1.3 Explain the Standard Setting Process

A description should be provided of what the standard setting is about, the particulars of the OER standard setting method, why the judges' participation is so critical in determining minimum passing scores, and examples should be given of how judges will carry out the OER standard setting process. As an example, a computerised presentation can be developed to cover all major points of the process. This presentation should be basic and should not assume any prior knowledge of the standard setting process from any of the participants. Questions should always be welcomed, and should be answered fully.

4.1.4 Provide Test Content & Rating Scale(s)

A well-defined table of specifications where the test content is clearly outlined needs to be provided in all situations. The rating scale for each extended response question should be provided and explained. This should not be the moment, though, for revisions or changes. Nevertheless, if a major flaw in the rating scale is identified, revisions before starting the OER standard setting process need to be made. The clarity of the rating scale is paramount to the reliability of the scoring and the OER standard setting process. Therefore, the scoring criteria for each item, and why it is so, must be clearly established before the start of the standard setting session.

4.1.5 Define Competency Levels

When determining minimum standards the judges need to understand the competency levels they will pass judgement on. In most licensure/certification programs, the minimum competency level to be evaluated is one. In these licensure/certification programs, the candidate either has the minimum requirements and passes, or does not meet these minimum requirements, and fails. These types of assessment programs require the judges to determine only one cut. Other programs where more distinctions in proficiency levels are needed may have several cut scores. Examples of such programs are: educational institutions that report exam results on multiple-grade scales (i.e., A-B-C-D-F), or assessment programs that report results on multiple performance levels (i.e., novice, apprentice, proficient, advanced). The following Conceptual Competency Graph provides a theoretical representation of different thresholds for a program with five competency levels. In this graph, each distribution shows the theoretical score ranges expected on a specific item, by a group of examinees typical of each competency category. It is worth noting that the normal score distributions typically observed on tests, derive from the conceptual application of the Central Limit Theorem to multiple tasks, for the population of examinees, across the full ability range. In this example, Highly Competent corresponds to a grade of A (top score) and Not Competent corresponds to a grade of F (failing score), for every score below the D cut score. In this scale, Marginal corresponds to a grade of C, which would indicate the examinee to be minimally competent but just passing. The conceptualisation represented in the graph clearly indicates that the thresholds to be used by the judges are not midpoints of the distributions of proficiencies of the students in each of the proficiency levels. Rather, they represent a homogeneous set of potential students at the absolute minimum level of proficiency that could be classified within each of the categories.

These students with the minimum level of proficiency for each category are described as the "borderline" examinees.

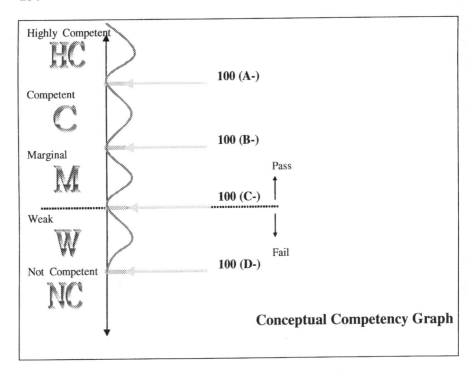

Figure 1. The Conceptual Competency Graph

At the beginning of the standard setting session, descriptions of "borderline" test-takers at each of the competency levels must be developed by the group. A "borderline" test-taker is an examinee whose knowledge and skills are at the borderline or lowest level (threshold) of each competency level. These definitions of "borderline" are quite critical and need to be arrived at in complete agreement by consensus of the standard setting judges. This process helps the group to integrate, establish common baselines, and get to work as a group. Example definitions of Highly Competent, Competent, and Marginal are presented below:

Table 1. Definitions of Mastery Levels

DEFINITIONS OF THRESHOLD MASTERY LEVELS

HIGHLY COMPETENT (minimal A)

Effectively communicates a **thorough** analysis and synthesis of all major concepts and themes in the question.
- Explains relationships among concepts.
- Provides evidence drawn from the course materials and/or the literature to support answer.
- There required by the question, chooses a position on an issue and provides rational justification for that position.
- Links theory to practice.

COMPETENT (minimal B)

Communicates effectively a **fundamental** analysis of most major concepts and themes in the question.
- Defines relationships among concepts.
- Provides some evidence drawn from the course to support the answer.
- Where required by the question, chooses a position on an issue and provides partial justification for that position.
- Where required by the question, links theory to practice.

MARGINAL (minimal C - Passing)

Communicates a **limited** understanding of the major concept(s) and theme(s) in the question.
- Identifies concepts and provides a superficial description of the relationship(s).
- Provides superficial/marginal evidence from the course to support answer.
- Where required by the question, chooses a position on an issue and provides discussion with marginal justification.
- When required by the question, superficially links theory to practice.

Table 2. Example Ratings of Item 1 Across Multiple Thresholds

RATER_____ **DATE**_____
#_____

CATEGORY: A - Highly Competent

ITEM	RATING POINTS					
	1	2	3	4	5	6
1	0	0	10	15	50	25

CATEGORY: B - Competent

ITEM	RATING POINTS					
	1	2	3	4	5	6
1	5	10	15	25	35	10

CATEGORY: C - Marginal

ITEM	RATING POINTS					
	1	2	3	4	5	6
1	10	20	40	20	10	0

4.1.6 Estimate Item Difficulty

Judges are instructed to estimate the number of 100 hypothetical students at the threshold of each of the competence categories [e.g.: Highly Competent (A), Competent (B), Marginal (C)] who would get distributed across all possible rating points for an item.
Example Instructions:

> "For each item estimate the percentage (or proportion) of 100 examinees in the specified category (A, B, or C) who would get each rating (1 to 6). Make sure that all 100 examinees are distributed across all possible points)".

As indicated by the ratings in Table 2, the expectation is that in higher competency levels the 100 examinees would be distributed more heavily at the higher rating points of the scale. As the level of competency decreases, the distribution of the 100 examinees becomes more heavily represented in mid and later at lower levels of the scale. It has been found that it is important to carefully check that judges' estimates add-up to 100 for each given distribution (Schmitt, 1999).

4.1.7   Reaching Informed Consensus

Another important element in the OER standard setting process is the process of reaching "informed consensus". After each item is rated across thresholds, judges are asked to verbally specify what their ratings were and to justify them. The process begins by asking judges for their individual rating across all levels of competency. The most extreme estimates are noted and those judges representing those viewpoints are asked to explain the reasons for their extreme ratings. The group is encouraged to not pre-judge the reasonableness of the estimates and to keep an open mind for the diverging explanations. After these points have been expounded, all judges are given a chance to revise their estimates based on the explanations given beforehand. In this way, "informed consensus" is achieved, and overall variability between ratings is minimised. If judges chose to remain discrepant, the different viewpoints are respected.

4.1.8   Maintain Parallel Standards

To maintain parallel standards across thresholds and items it is important to ascertain the reasonableness of the distributions for the same item across different thresholds and of different items within a threshold. It is an important check of this method, that this "reasonableness test" should apply both across thresholds for the same item for the same judge, and also across items for each threshold, also for the same judge, and across judges. This "reasonableness" check is achieved by having the judges evaluate their ratings across different thresholds and items. The panel facilitator addresses discrepancies within each judge's ratings, as well as those observed between members of the panel. The process involves active interaction between all participants, examining the justifications for each significantly discrepant score. The end result of the process is a new consensus which might or might not eliminate the discrepancies, but which has certainly examined and provided a basis for any possible remaining differences or adjustments carried out. Remaining differences are where the multidimensionality of the scale can be correctly represented and adjusted for, making it possible for a final score to represent dimensional differences across items and thresholds.

4.1.9 Setting Cut Scores for the Item and Test

The probability distributions resulting from this procedure are used to set the cut scores for each examination by averaging the judges' ratings across items at each competency level. An example of a summary page identifying cut scores for a test for three different thresholds is presented in Table 3. The average cut score point for each of four items is given at the far right column and the overall cut score is presented as either a cut score or percent correct across all four items and across all eight judges. In cases where discrepancies in standards across judges are determined to be too large, the rating for the outlier judge can be deleted and averages computed again with the remaining data.

Table 3. Standard Setting Score by Category (A, B, C)

CATEGORY: A - Highly Competent

ITEM #	RATERS								AVERAGE ALL
	#1	#2	#3	#4	#5	#6	#7	#8	
1	5	6	6	6	6	6	6	6	6
2	5	6	5	5	5	5	5	5	5
3	6	5	6	4	6	5	4	4	5
4	4	4	4	5	5	4	6	6	5
Passing Score	5	5	5	5	6	5	5	5	**5**
Passing Percent	83%	88%	88%	83%	92%	83%	88%	88%	**86%**

CATEGORY: B - Competent

ITEM #	PANELISTS								AVERAGE ALL
	#1	#2	#3	#4	#5	#6	#7	#8	
1	4	5	5	5	5	5	5	5	5
2	3	5	4	4	4	4	4	4	4
3	4	4	5	3	5	4	3	3	4
4	4	3	3	4	4	3	5	5	4
Passing Score	4	4	4	4	5	4	4	4	**4**
Passing Percent	63%	71%	71%	67%	75%	67%	71%	71%	**69%**

CATEGORY: C - Marginal

ITEM #	PANELISTS								AVERAGE ALL
	#1	#2	#3	#4	#5	#6	#7	#8	
1	3	4	4	4	4	4	4	4	4
2	2	4	3	3	3	3	3	3	3
3	3	3	4	2	4	3	2	2	3
4	3	2	2	3	3	2	4	4	3
Passing Score	3	3	3	3	4	3	3	3	**3**
Passing Percent	46%	54%	54%	50%	58%	50%	54%	54%	**53%**

4.2 Application of the Optimised Extended Response Standard Setting Method

The OER Standard Setting Method has been successfully implemented to higher education examinations, at the US graduate and undergraduate levels, with constructed-response items that involved extended writing in response to several prompts (Schmitt, 1999). These examinations had several items, many of them with different scales. The OER Standard Setting Method proved easy to explain to panellists/judges and flexible and accurate for the use with the different rating scales. Inter-judge reliability estimates (phi-coefficient) ranged between .80 and .94. Raymond and Reid (2001) consider desirable coefficients of .80 and greater.

4.2.1 Outcome of Examinations

The exams used as examples in this study were used to grant three semester hours of upper-level undergraduate credit to students who receive a score equivalent to a letter-grade of C or higher on the examination. An example of the table used to record the OER Standard Setting Method across thresholds is presented in Table 4. This Table could be used as a model for an implementation of the OER Standard Setting Method.

Table 4. Example of a form to record the OER Standard Setting across thresholds

RATER_____   DATE_____
#_____

CATEGORY: A - Highly Competent

ITEM	RATINGS					
	1	2	3	4	5	6
1						
2						
3						
4						
5						

CATEGORY: B - Competent

ITEM	RATINGS					
	1	2	3	4	5	6
1						
2						
3						
4						
5						

CATEGORY: C - Marginal

ITEM	RATINGS					
	1	2	3	4	5	6
1						
2						
3						
4						
5						

For each item estimate the percentage (or proportion) of 100 examinees in the specified category (A, B, or C) who would get each rating.

5. DISCUSSION

The OER Standard Setting Method discussed presents a valuable option in order to deal with the multidimensional scales that are found in extended response examinations. As originally developed (Schmitt, 1999) this method's well defined rating scales provide a reliable procedure to determine the different scoring points where judges estimate minimum passing points for each scale, which has been shown to work in various examinations in different content areas in traditional paper-and-pencil administrations, as well as in computer delivered examinations (Cascallar, 2000). As such it addresses the challenge presented by the many complexities in the application of standards to constructed-response tests in a variety of settings and forms of administration.

Recent conceptualisations, such as those differentiating between criterion- and construct-referenced assessments (William, 1997), present very interesting distinctions between the descriptions of levels and the domains. This method can integrate the conceptualisation, as providing both an adequate "description" of the levels, as attained by the consensus of the judges, as well as a flexible "exemplification" of the level inherent in the process to reach the consensus. As it has been pointed out, there is an essential need to estimate the procedural validity (Hambleton, Jaeger, Plake, & Mills, 2000; Kane, 1994) of judgement-based cutoff scores. This line of research will eventually lead to the most desirable techniques to guide the judges in providing their estimates of probability. In this endeavour the OER Standard Setting Method suggests a methodology and provides the procedures to maintain the necessary degree of consistency to make critical decisions that affect examinees in the different settings in which their performance is measured against the cut scores set using standard setting procedures. With reliability being a necessary but not sufficient condition for validity, it is necessary to investigate and establish valid methods for the setting of those cutoff points (Plake & Impara, 1996).

The general uneasiness with the current standard setting methods (Pellegrino, Jones, & Mitchell, 1999) rests to a great extent on the fact that setting standards is a judgement process that needs well-defined procedures, well-prepared judges, and the corresponding validity evidence. validity evidence is essential to reach the quality commensurate with the importance of its application in many settings (Hambleton, 2001). Ultimately, the setting of standards is a question of values and of the decision-making involved in the evaluation of the relative weight of the two types of errors of classification (Zieky, 2001). As there are no purely absolute standards, and problems are identified with various existing methods (Hambleton, 2001), it is imperative to remember and heed the often-cited words by Ebel (1972), which are still current today:

> Anyone who expects to discover the "real" passing score by any of these approaches, or any other approach, is doomed to disappointment, for a "real" passing score does not exist to be discovered. All any examining authority that must set passing scores can hope for, and all any of their examinees can ask, is that the basis for defining the passing score be defined clearly, and that the definition be as rational as possible. (p. 496)

It is expected that the OER Standard Setting Method will provide a better way to determine passing scores for extended response examinations where multidimensionality could be an issue and in this way provide a framework to more accurately capture the elements leading to quality standard setting processes, and ultimately to more reliable, fairer, and valid evaluation of knowledge.

REFERENCES

American Educational Research Association, American Psychological Association, & National Council on Measurement in Education (1999). *Standards for educational and psychological testing.* Washington, DC: American Psychological Association.

Angoff, W. H. (1971). Scales, norms, and equivalent scores. In R. L. Thorndike (Ed.), *Educational measurement* (2nd ed.) (pp. 508-600). Washington, DC: American Council on Education.

Cascallar, A. S. (2000). *Regents College Examinations.* Technical Handbook. Albany, NY: Regents College.

Cizek, G. J. (2001). Conjectures on the rise and call of standard setting: An introduction to context and practice. In G. J. Cizek (Ed.), *Setting performance standards: Concepts, methods, and perspectives.* Mahwah, NJ: Lawrence Erlbaum Publishers.

Cohen, A. S., Kane, M. T., & Crooks, T. J. (1999). A generalized examinee-centered method for setting standards on achievement tests. *Applied Measurement in Education, 12*, 343-366.

Ebel, R. L. (1965). *Measuring educational achievement.* Englewood Cliffs, NJ: Prentice-Hall.

Ebel, R. L. (1972). *Essentials of educational measurement.* (2nd ed.) Englewood Cliffs, NJ: Prentice-Hall.

Faggen, J. (1994). *Setting standards for constructed response tests: An overview.* Princeton, NJ: Educational Testing Service.

Fitzpatrick, A. R. (1989). Social influences in standard-setting: The effects of social interaction on group judgments. *Review of Educational Research, 59*, 315-328.

Flanagan, J. C. (1951). Units, scores and norms. In E. F. Lindquist (Ed.), *Educational Measurement* (pp. 695-763). Washington, DC: American Council on Education.

Glaser, R. (1963). Instructional technology and the measurement of learning outcomes. *American Psychologist, 18*, 519-521.

Glaser. R., & Klaus, D. J. (1962). Proficiency measurement: Assessing human performance. In R. M. Gagne (Ed.), *Psychological principles in systems development.* New York: Holt, Rinehart, and Winston.

Hambleton, R. K. (2001). Setting performance standards on educational assessments and criteria for evaluating the process. In G. J. Cizek (Ed.), *Setting performance standards: Concepts, methods, and perspectives.* Mahwah, NJ: Lawrence Erlbaum Publishers.

Hambleton, R. K., & Novick, M. R. (1972). *Toward an integration of theory and method for criterion-referenced tests.* Iowa City: The American College Testing Program.

Hambleton, R. K., Jaeger, R. M., Plake, B. S., & Mills, C. N. (2000). Setting performance standards on complex educational assessments. *Applied Psychological Measurement, 24* (4), 355-366.

Harvey, A. L., & Way, W. D. (1999). *A comparison of web-based standard setting and monitored standard setting.* Paper presented at the annual meeting of the National Council on Measurement in Education. Montreal, Canada.

Hurtz, G. M., & Hertz, N. R. (1999). How many raters should be used for establishing cutoff scores with the Angoff method? A generalizability theory study. *Educational and Psychological Measurement, 59,* 885-897.

Jaeger, R. M. (1989). Certification of student competence. In R. L. Lynn (Ed.), *Educational Measurement* (3rd ed., pp. 485-514). Washington, DC: American Council on Education.

Jaeger, R. M. (1994) *Setting performance standards through two-stage judgmental policy capturing.* Presented at the annual meetings of the American Educational Research Association and the National Council on Measurement in Education, New Orleans.

Kane, M. (1994). Validating the performance standards associated with passing scores. *Review of Educational Research, 64,* 425-462.

Kane, M. (2001). So much remains the same: Conception and status of validation in setting standards. In G. J. Cizek (Ed.), *Setting performance standards: Concepts, methods, and perspectives.* Mahwah, NJ: Lawrence Erlbaum Publishers.

Maurer, T. J., Alexander, R. A., Callahan, C. M., Bailey, J. J., & Dambrot, F. H. (1991). Methodological and psychometric issues in setting cutoff scores using the Angoff method. *Personnel Psychology, 44,* 235-262.

Mehrens, W. A., & Cizek, G. J. (2001). Standard setting and the public good: Benefits accrued and anticipated. In G. J. Cizek (Ed.), *Setting performance standards: Concepts, methods, and perspectives.* Mahwah, NJ: Lawrence Erlbaum Publishers.

Millman, J. (1973). Passing scores and test lengths for domain-referenced measures. *Review of Educational Research, 43,* 205-217.

Norcini, J. J., Shea, J., & Grosso, L. (1991). The effect of numbers of experts and common items on cutting score equivalents based on expert judgment. *Applied Psychological Measurement, 15,* 241-246.

Pellegrino, J. W , Jones, L. R., & Mitchell, K. J. (Eds.). (1999). *Grading the nation's report card.* Washington, DC: National Academy Press.

Plake, B. S. (1998). Setting performance standards for professional licensure and certification. *Applied Measurement in Education, 11,* 65-80.

Plake, B. S., & Impara, J. C. (1996). *Intrajudge consistency using the Angoff standard setting method.* Paper presented at the annual meeting of the National Council on Measurement in Education. New York, NY.

Plake, B. S., Melican, G. M., & Mills. C. N. (1991). Factors influencing intrajudge consistency during standard-setting. *Educational Measurement: Issues and Practice, 10,* 15-16, 22, 25-26.

Plake, B. S., Impara, J. C., & Irwin, P. M. (2000). Consistency of Angoff-based predictions of item performance: Evidence of technical quality of results from the Angoff standard setting method. *Journal of Educational Measurement, 37,* 347-355.

Raymond, M. R., & Reid, J. B. (2001). Who made thee a judge? Selecting and training participants for standard setting. In G. J. Cizek (Ed.), *Setting performance standards: Concepts, methods, and perspectives.* Mahwah, NJ: Lawrence Erlbaum Publishers.

Schmitt, A. (1999). *The Optimized Extended Response Standard Setting Method.* Technical Report, Psychometric Division. Albany, NY: Regents College.

Shepard, L. A. (1980). Standard-setting issues and methods. *Applied Psychological Measurement, 4,* 447-467.

Sireci, S. G., & Biskin, G. H. (1992). Measurement practices in national licensing examination programs: A survey. *Clear Exam Review, 3* (1), 21-25.

Sireci, S. G., Robin, F., & Patelis, T. (1999). Using cluster analysis to facilitate standard setting. *Applied Measurement in Education, 12*, 301-325.

William, D. (1997). *Construct-referenced assessment of authentic tasks: alternatives to norms and criteria.* Paper presented at the 7th Conference of the European Association for Research in Learning and Instruction. Athens, Greece. August 26-30.

Zieky, M. J. (2001). So much has changed: How the setting of cutscores has evolved since the 1980's. In G. J. Cizek (Ed.), *Setting performance standards: Concepts, methods, and perspectives.* Mahwah, NJ: Lawrence Erlbaum Publishers.

Assessment and Technology

Henry Braun
Educational Testing Service, Princeton New Jersey, USA

1. INTRODUCTION

Formal assessment has always been a critical component of the educational process, determining entrance, promotion and graduation. In many countries (e.g. United States, England & Wales, Australia) assessment has assumed an even more salient place in governmental policy over the last two decades. Attention has focused on assessment because of its role in monitoring system functioning at different levels (students, teachers, schools and districts) for purposes of accountability and, potentially, spearheading school improvement efforts. At the same time, new information technologies have exploded on the world scene with enormous impact on all sectors of the economy, education not excepted.

In the case of pre-college education, however, change (technological and otherwise) has come mostly at the margins. The core functions in most educational systems have not been much affected. The reasons include the pace at which technology has been introduced into schools, the organisation of technology resources (e.g. computer labs), poor technical support and lack of appropriate professional development for teachers. Accordingly, schools are more likely to introduce applications courses (e.g. word processing, spreadsheets) or "drill-and-kill" activities rather than finding imaginative ways of incorporating technology into classroom practice. While there are certainly many fine examples of using technology to enhance motivation and improve learning, very few have been scaled up to an appreciable degree.

Notwithstanding the above, it is likely that the convergence of computers, multimedia and powerful communication networks will eventually leave their mark on the world of education, and on assessment in particular. Certainly, technology has the potential to increase the value of and enhance access to assessment. It can also improve the efficiency of assessment processes. In this chapter, we present and explicate a structure for exploring the relationship between assessment and technology. While the structure should apply quite generally, we confine our discussion primarily to U.S. pre-college education.

In our analysis, we distinguish between direct and indirect effects of technology. By the former we refer to the tools and affordances that change the practice of assessment and are the principal focus of attention in the research literature. Excellent examples are provided by Bennett (1998); Bennett (2001) and Bunderson, Inouye, and Olsen, (1989). In these studies the authors project how the exponential increase in available computing power and the advent of affordable high speed data networks will affect the design and delivery of tests, lead to novel features and, ultimately, to powerful new assessment systems that are more tightly coupled to instruction.

There is noticeably less attention to what we might term the indirect effects of technology; that is, how technology helps to shape the political-economic context and market environment in which decisions about assessment take place. These decisions, concerning priorities and resource allocation, exert considerable influence on the evolution of assessment. Indeed, one can argue that while science and technology give rise to an infinite variety of possible assessment futures, it is the forces at play in the larger environment that determine which of these futures is actually realised. For this reason, it is important for educators to appreciate the different ways in which technology can and will influence assessment. With a deeper understanding of this relationship, they will be better prepared to help society harness technology in ways that are educationally productive. Bennett (2002) offers a closely related analysis of the relationship between assessment and technology, with a detailed discussion of current developments in the U.S. along with informed speculation about the future.

Section 2 begins by presenting a framework for the study of assessment. In Section 3 we explore the direct effects of technology, followed in Section 4 by an analysis of how technology can contribute to assessment quality. Section 5 discusses the indirect effects of technology and Section 6 the relationship between assessment purpose and technology. The final Section 7 offers some conclusions.

2. A FRAMEWORK FOR ANALYSIS

Braun (2000) proposed a framework to facilitate the analysis of forces like technology, which shape the practice of assessment. The framework comprises three dimensions: Context, Purpose and Assets. (See Figure 1.)

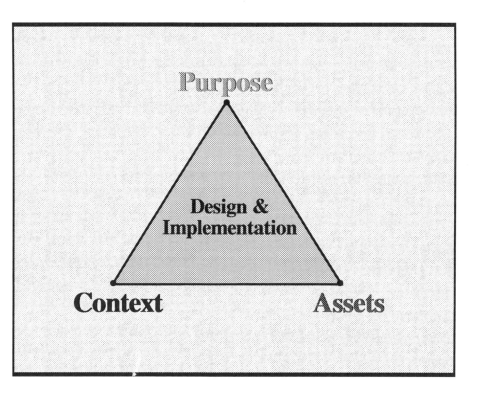

Figure 1. Three dimensions of assessment

Context refers to: (i) the physical, cultural or virtual environment in which assessment takes place; (ii) the providers and consumers of assessment services; and (iii) any relevant political and economic considerations. (See Figure 2.) For example, the microenvironment may range from a typical fourth grade classroom in a traditional public school to the bedroom of a home-schooled high school student taking an online physics course.

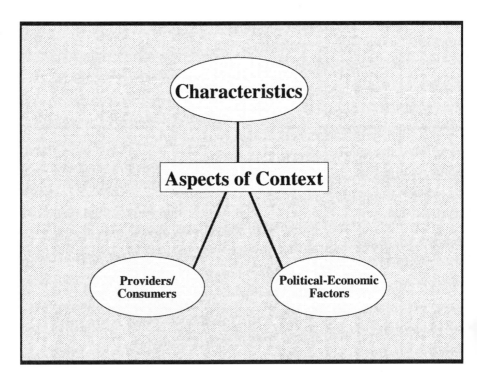

Figure 2. The context dimension of assessment

In the former case, the providers range from the teacher administering an in-class test to the publisher of a standardised end-of-year assessment to be used for purposes of accountability. For the in-class test, the student and the teacher are the primary consumers, while for the end-of-year assessment the primary consumers are the school and government officials as well as the public at large. In the latter case, the provider is likely to be some combination of a for-profit company and the school faculty while the primary consumer is the student and her parent.

The macroenvironment is largely characterised by political and economic considerations. The rewards and sanctions (if any) attached to the end-of-year assessment, along with the funding allocated to it, will shape the assessment program and its impact on classroom activities. In the online course, institutional interest in both reducing student attrition and establishing the credibility of the program will influence the nature of the assessments employed.

The second dimension, purpose, also has three aspects: Choose, Learn and Qualify. (See Figure 3.) The data from an assessment can be used to choose a program of study or a particular course within a program. Other assessments serve learning by providing information that can be used by the student to track progress or diagnose strengths and weaknesses. Finally assessments can determine whether the student obtains a certificate or other qualification that enables them to attain their goals. Although these purposes are quite distinct, a single assessment may well serve multiple purposes. For example, results from a selection test can sometimes be used to guide instruction, while a portfolio of student work culled from assessments conducted during a course of study can inform a decision about whether the student should receive a passing grade or a certificate of completion.

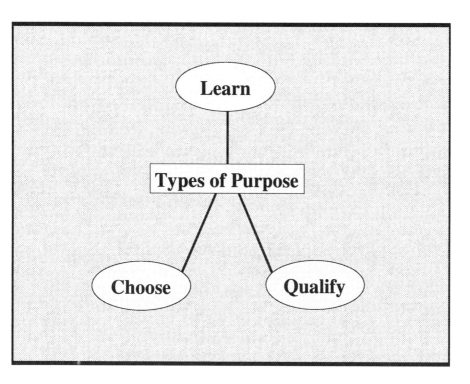

Figure 3. The purpose dimension of assessment

In classroom settings, external tests (alone or in conjunction with classroom performance) are sometimes used for tracking purposes. Teachers will employ informal assessments during the course of the year to inform instruction and may subscribe to services offered by commercial firms in order to enable their students to practice for the end-of-year assessment, which relates to the "Qualify" aspect of purpose. Typically, governmental initiatives in assessment focus on this third aspect with the aim of establishing a basis for accountability.

In the case of online learning, students may employ preliminary assessments to decide if they are ready to enter the program. Later they will use assessments to monitor their progress in different classes and subsequently sit for course exams to determine whether they have passed or failed.

The third dimension, assets, represents what developers bring to bear on the design, development and implementation of an assessment. It consists of three components: Disciplinary knowledge, cognitive/measurement science and infrastructure. (See Figure 4.) The first refers to the subject matter knowledge (declarative, procedural and conceptual) that is the focus of instruction. The second component refers to the understandings, models and methods of both cognitive science and measurement science that are relevant to test construction and test analysis. Finally, infrastructure comprises the systems of production and use that support the assessment program. These systems include the hardware, software, tools and databases that are needed to carry out the work of the program.

In the following sections we will examine the effects of technology on assessment. We will be concerned with not only whether a particular set of technologies has or can have an impact on assessment practice but also its contribution to assessment quality. How is quality defined? We posit that assessment quality has two essential aspects, denoted by validity and efficiency.

The term validity encompasses psychometric characteristics (i.e. accuracy and reliability), construct validity (i.e. whether the test actually measures what it purports to measure) and systemic validity (i.e. its effect on the educational system). Efficiency refers to the monetary cost and time involved in production, administration and reporting. While cost and time are usually closely linked, they are sometimes distinct enough to warrant separate consideration. As we shall see below, validity and efficiency often represent countervailing considerations in assessment design, with the balance point determined by both context and purpose.

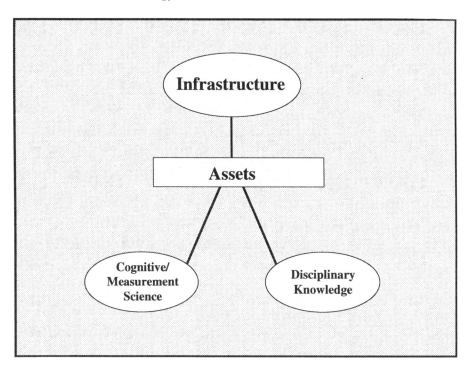

Figure 4. The assets dimension of assessment

Indeed, quality is a multidimensional construct, which can legitimately be viewed differently by different stakeholders. For example, governmental decision makers often give primacy to the demands of efficiency over considerations of validity, particularly when the "more valid" solutions are not readily available or require longer time horizons. Educators, on the other hand, typically focus on validity concerns although they are certainly not indifferent to issues of cost and time. These conflicting views play out over time in many different settings.

3. THE DIRECT EFFECTS OF TECHNOLOGY

In view of the definition of the direct effects of technology offered in the introduction, they are best studied by consideration of the infrastructure component of the Assets dimension. To appreciate the impact of technology, we require, at least at a schematic level, a model of the process for assessment design and implementation. This is presented in Figure 5.

Somewhat different and more elaborated models can be found in Bachman and Palmer (1996) and Mislevy, Steinberg, and Almond (2002).

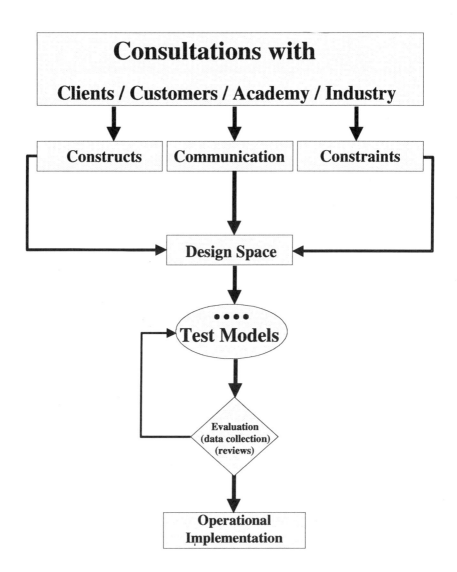

Figure 5. A model of the process for assessment design and implementation.

The first phase of the process leads to the identification of the design space; that is, the set of feasible assessment designs. The design space is determined by the "three C's": Constructs, Claims and Constraints. Constructs are the targets of inference drawn from the substantive field in question, while claims are the operational specifications of those inference targets expressed in terms of what the students know or can do. Constraints are the limitations (physical, monetary, temporal) that must be taken into account by any acceptable design.

Once the boundaries of the design space are known, different designs can be generated, examined and revised through a number of cycles, until a final design is obtained. (At this stage, operational issues can also be addressed.) With a putative final design in hand, construction of the instrument can begin. In practice, this work is usually embedded in a larger systems development effort that supports subsequent activities including test administration, analysis and reporting.

Technology may exert its influence at all phases of the process. For example, the transition from constructs to claims can be very time consuming, often requiring substantial knowledge engineering. Shute and her collaborators (Shute, Torreano & Willis, 2000; Shute & Torreano, 2001) have developed software that, in preliminary trials, has markedly reduced the time required to elicit and organise expert knowledge for both assessment and instructional purposes. Future versions should yield improved coverage and fidelity of the claims as well, resulting in both increased efficiency and enhanced validity.

Shifting attention to the implementation phases, technology makes possible tools that result in the automation, standardisation and enhancement of different assessment processes, rendering them more efficient. Increased efficiency, in turn, yields a greater number of feasible designs, some of which may yield greater validity. Item development and the scoring of constructed responses illustrate the point.

One of the more time consuming tasks in the assessment process is the development of test items. Even in testing organisations with large cadres of experienced developers, creating vast pools of items meeting rigorous specifications is an expensive undertaking. Over the last five years, tools for assisting developers in item generation have come into use, improving efficiency by factors of 10 to 20. In the near future, these tools will be enhanced so that developers can build items with specified psychometric characteristics. This will result in further gains in efficiency as well as some improvement in the quality of the item pools. Eventually, item libraries will consist of shells or templates with items generated on demand by selecting a specific template along with the appropriate parameters (Bejar, 2002). Beyond further improvements in efficiency, item generation on demand has implications both for the security of high stakes tests and the feasibility of offering customised practice tests in instructional settings.

Test questions that require the student to produce a response are often regarded as more desirable than multiple choice questions (Mitchell, 1992). Grading the responses, however, imposes administrative and financial burdens that are often prohibitive. Arguably, the dominance of the multiple-choice format is due in large part to its cost advantage as well as its contribution to score reliability. In the early 90's, the advent of computer-delivered tests heralded the possibility of reducing reliance on multiple choice items. It became apparent that it would be necessary to develop systems to automatically score student constructed responses. In fact, expert systems to analyse and evaluate even complex products such as architectural drawings and prose essays as well as mathematical expressions have been put into operation (Bejar & Braun, 1999; Burstein, Wolff, and Lu, (1999).

Automated scoring yields substantial improvements in cost and time over human scoring, usually with no diminution accuracy. In the course of developing such systems, a more rigorous approach to both question development and response scoring proves essential and yields, as an ancillary benefit, modest improvements in test validity. This argument was developed by Bejar and Braun (1994) in the context of architectural licensure but holds more generally.

Computer delivery is perhaps the most obvious application of technology to assessment, making possible a host of innovations including the presentation of multimedia stimulus materials and the recording of responses in different sensory modalities. (See Bennett (1998) for further discussion.) In conjunction with automated scoring capabilities, this makes practical a broad range of performance assessments.

With sufficient computing power and fast networks, various levels of interactivity are possible. These range from adaptive testing algorithms (Swanson and Stocking, 1993) to implementations of Bayes inference networks that can dynamically update cognitively grounded student profiles and suggest appropriate tasks or activities for follow up (Mislevy, Almond & Steinberg, 1999). The paradigmatic applications are the complex interactive simulations or intelligent tutoring systems that have been developed by such companies as Maxis and Cognitive Arts for the entertainment or education/training markets.

The power of technology is magnified when individual tools are organised into coherent combinations that can accomplish multiple tasks. Gains in efficiency can then be realised both at the task and the system levels. An early instance is the infrastructure that was built to develop and deliver computer-based architectural simulations that are part of a battery of tests used for the registration (licensure) of architects in the U.S. and Canada (Bejar & Braun, 1999). The battery includes fifteen different types of simulations, each one intended to elicit from candidates a variety of complex graphical responses. Furthermore, to meet the client's targets for cost savings, these responses have to be scored automatically without human intervention.

The core of the infrastructure comprises a set of interrelated tools that supports authoring of multiple instances of each simulation type, efficient rendering of the geometric objects for each such instance, delivery of the simulations, capture of the candidates' data as well as the analysis and evaluation of the responses. In comparison to the previous paper-and-pencil system, the current system achieves greater fidelity with respect to the constructs of interest and more uniformity in scoring accuracy, while possessing superior psychometric properties. (The latter include better comparability of test forms over time and higher reliability of classification decisions.) It also offers candidates significantly more opportunities to sit for the test and much more rapid reporting of results. This situation stands in stark contrast with most high stakes paper-and-pencil assessments (particularly those incorporating some constructed response questions), which are offered only once or twice a year with results reported months after the administration.

A later and more sophisticated example is the infrastructure developed by Mislevy and his associates (Mislevy, Steinberg, Breyer, Almond & Johnson, 1999) to facilitate the implementation of a general assessment design process, evidence centred design. The infrastructure comprises both design objects and delivery process components. Design objects are employed to build student models and task models. In addition there are systems to extract evidence from student work and to dynamically update the student model on the basis of that evidence. Finally, there are software components that support task selection, task presentation, evidence identification and evidence accumulation. The idea is to build flexible modular systems with reusable components that can be configured in different ways to efficiently support a wide variety of assessment designs.

4. ASSESSMENT QUALITY

How does a technology-driven infrastructure contribute to the quality of an assessment system? Recall that we asserted that quality comprises two aspects, validity and efficiency. In some settings, we can improve efficiency but with little effect on validity. This is illustrated by the use of item generation tools to reduce the cost of developing multiple choice items for tests of fixed design. In other settings, we can enhance validity but with little effect on efficiency. An obvious case in point is lengthening a test by the addition of items in order to gain better coverage of the target constructs.

Validity is improved but at the price of some additional cost and testing time.

Improvements to efficiency or validity (but not both) are typical of the choices that test designers must make. That is, there is a trade-off between validity and efficiency. Consider that for reasons of coverage and fairness, test designs employing different item types are generally preferred to those that only use multiple choice items. However, incorporating performance items generally raises costs by an order of magnitude and therefore tends to preclude their inclusion in the final design. Such considerations are often paramount in public school settings, where the cost of hand scoring many thousands of essays or mathematical problems, as well as the associated time delays, are quite burdensome. Indeed, under the pressure of increased testing for accountability, the states of Florida and Maryland recently announced that they were reducing or eliminating the use of performance assessments in their end-of-year testing programs.

Technology can make a critical contribution by facilitating the attainment of a more satisfactory balance point between validity and efficiency. This parallels the notion that technology can render obsolete the traditional trade-off between "richness and reach". Popular in some of the recent business literature (Evans & Wurster, 2000), the argument is that in the past one had to choose between offering a rich experience to a select few and delivering a comparatively thin experience to a larger group. Imagine, for example, the difference between watching a play in the theatre and reading it in a book. Through video technology a much larger audience can watch the play, enjoying a much richer experience than that of the reader --- though perhaps not quite as rich as that of the playgoer. The promise of the new, technology-mediated trade-off between richness and reach was responsible (at least in part) for the explosion of investment in education-related internet companies. The subsequent collapse of the "dot.com bubble" does not render invalid the seminal insight but, rather, is more a reflection of the difficulty in creating new markets and the unavoidable consequence of "irrational exuberance".

In the realm of assessment, the contribution of technology stems from its potential to substantially enhance the efficiency of certain key processes. This, in turn, can dramatically increase the set of feasible designs, resulting in a final design of greater validity. Two striking examples are the automated scoring of (complex) student responses and the implementation of adaptive assessment. Consider the following examples.

A number of essay grading systems are now available, including e-rater (Burstein et al.,1999) and KAT (Wolfe, Schreiner,, Rehder, Laham, Foltz, & Landauer, 1998). Although based on different methodologies, they are all able to carry out the analysis and evaluation of student essays of several hundred words. In the case of e-rater, the graded essays are responses to one of a set of pre-determined prompts on which the system has been trained. In its current version, e-rater not only provides scale scores and general feedback but also detailed diagnostics linked to the individual student's writing.

A computer-based adaptive mathematics assessment, ALEKS, developed by Falmagne and associates (Falmagne, Doignon, Koppen, Villano, & Johannesen, 1990; See also http://www.aleks.com), is able in a relatively short time to place a student along a continuum of development in pre-collegiate mathematics curriculum. Based on many years of research in cognitive psychology and mathematics education, ALEKS enables a teacher or counsellor to direct the student to an appropriate class and supports the setting of initial instructional goals.

Intelligent tutoring systems (Snow & Mandinach, 1999; Wenger, 1987) are intended to provide adaptive support to learners. They have been built for a variety of content areas such as geometry (Anderson, Boyle, & Yost, 1985), computer programming (Anderson, & Reiser, 1984) as well as electronic troubleshooting (Lesgold, Lajoie, Bunzo, & Egan, 1992). A related example at ETS is Hydrive (Mislevy & Gitomer, 1996) which was built to support the training of flight-line mechanics on the hydraulic systems of the F-15. It enables the trainees to develop and hone their skills on a large library of troubleshooting problems. They are given wide scope in how to approach each problem as well as a choice in the level and type of feedback. Although they are not aware of it, the feedback is based on a comprehensive cognitive analysis of the domain and a sophisticated psychometric model (matched to the cognitive analysis), which is dynamically updated with each action the student takes.

Intelligent tutoring systems in the workplace can be very efficient in terms of the utilisation of expensive equipment. Apprentices can be required to meet certain performance standards before being allowed to work on actual equipment, such as jet aircraft. When they do, they are much more likely to profit from the experience. Suppose, in addition, that the problem library is constantly updated to reflect the problems faced by experts as new or modified equipment comes on line. Students then have the benefit of being trained on exactly the sort of work they will be expected to do when they graduate, largely eliminating the notorious transfer of training problem. Such a seamless transition from training to work is exceptionally rare, at least in the U.S.

What these examples have in common is that they illustrate how technology can provide large populations of learners with access to assessment services that, until recently, were only available to the few students blessed with exceptional teachers. In fact, for some purposes, these systems (or ones available shortly) are not surpassed by even the best teachers. Consider that, with e-rater, students can write essentially as many essays as they like, revise them as often as they want, whenever they wish --- and each time receive immediate, detailed feedback. Moreover, it is possible to receive scale scores based on different sets of standards, provided that the system has been trained to those standards. High school students could then be graded according to the standards set by the English teachers in their school, all English teachers in their system or even those set by the English teachers at the state college that many hope to attend. The juxtaposition of these scores can be instructive for the student and, in the aggregate, serve as an impetus for professional development for the secondary school teachers.

If these developments continue to ramify (depending, in part, on sufficient investments), technology will emerge as a force for the democratisation of assessment. It will facilitate the use of more valid assessments in a broader range of settings than heretofore possible. This increase in validity will usually be obtained with no loss of efficiency and, perhaps, even with gains in efficiency if a sufficiently long time horizon is used.

5. INDIRECT EFFECTS OF TECHNOLOGY

Our analysis begins by returning to the framework of Figure 4. While the discussion in Section 3 focused on the infrastructure aspect of the Assets dimension, consideration of the indirect effects of technology begins with the other two aspects of the Assets dimension.

It is evident that technology – in the form of tool sets and computing power – has an enormous effect on the development of many disciplines, especially the sciences and engineering. To cite but two recent examples: (i) The development of machines capable of rapid gene sequencing were critical to the successful completion of the Human Genome Project; (ii) The Hubble Space Telescope has revealed unprecedented details of familiar celestial objects as well as entirely new ones that have led to advances in cosmology. Experimental breakthroughs lead, over time, to reconceptualisations of a discipline and, ultimately, to new targets for assessment.

Similarly, developments in cognitive science, particularly in understanding how people learn, gradually influence the constructs and models that impact the design of assessments for learning (Pellegrino, Chudowski, & Glaser, 2001). Again, these advances depend in part on imaging technology, though this is not the major impetus for the evolution of the field. However, these developments do influence measurement models that are proposed to capture the salient features of these learning theories. Of course, technology plays a critical role in making these more complex measurement models practical --- but this brings us back to technology's direct effects! In short, through its impact on the different disciplines, technology influences the constructs, claims and models that, in turn, shape the practice of assessment.

Returning now to the framework of Figure 2, we examine the indirect effects of technology through the prism of Context. There is a complex interplay between technology on the one hand and political-economic forces on the other. One view is that new technologies can stimulate political and economic actions that, in turn, influence the educational environment. In most countries, the prospect of increasing economic competitiveness has spurred considerable governmental investment in computers and communication technology for schools. In the U.S., for example, over the last decade states have made considerable progress toward the goal of substantially reducing the pupil-computer ratio. In addition, the U.S. government through its E-rate program has subsidised the establishment of internet connections for public schools. Indeed, both at the state and federal levels, there is the hope that the aggressive pursuit of a technology in education policy can begin to equalise resources and, eventually, achievement across schools. That is, again, that technology can act as a democratising force in education.

We have already made the point that the promise of technology to effect a new trade-off between richness and reach fuelled interest in internet-based education companies. Hundreds of such companies were founded and attracted, in the aggregate, billions of dollars in capital. While many offered innovative products and services, few were successful in establishing a viable business. Now that most of those companies have failed or faltered and been bought out, we are left with just a handful of behemoths astride the landscape.

One consequence is that in the U.S. four multinational publishers dominate the education market with ambitious plans to provide integrated suites of products and services to their customers. If they are successful, they will gain unprecedented influence over the conduct of pre-college education. In particular, assessment offerings are much more likely to be determined by financial calculations based on the entire suite to be offered. For example, a firm with a large library of multiple choice items on hand will naturally want to leverage that asset by offering those same items in an electronic format. Moreover, it will be able to offer the prospective buyer much more favourable terms than would be the case if it had to build a new pool of items incorporating performance assessment tasks and the accompanying scoring systems. In the U.S., at least, these conservative tendencies will be reinforced by the pressure to deliver practice tests for the high stakes outcome assessments that are now mandated by law.

Thus, technology interacts with the political, economic and market forces that help to shape the environments in which assessment takes place. In the case of instructional assessment, the presence of an internet connection in a classroom expands the range of such assessments that can be delivered to the students. For the moment, the impact on the individual student is limited both by the amount of time they can actually access the internet and the quality of the materials available. The first issue, which we can term "effective connectivity", will diminish with the continuing expansion of traditional computer resources and the increasing penetration of wireless technology, The latter issue is more problematic. In U.S. public education systems there is considerable pressure to focus such assessment activities on preparing students to take end-of-year tests tied to state-level accountability. Consequently, they tend to rely on standard multiple-choice items that target basic competencies.

In the case of assessment for other purposes, the situation is somewhat different. The existence of a network of testing centres with secure electronic transmission capabilities makes possible the delivery of high stakes computer-based tests for both selection and qualification. Examples of the former (for college and beyond) are the GRE, GMAT and TOEFL. Examples (in the U.S.) of the latter are tests for professional licensure in medicine and architecture. Internationally, certifications in various IT specialities are routinely delivered on computer. It should be noted, however, that with the notable exceptions of medicine and architecture, the formats and content of these assessments have not changed much with computer delivery. High stakes assessments are subject to many (conservative) forces that appear to make real change difficult to accomplish.

Technology can shape the design space by contributing to the creation of novel learning environments (e.g. e-learning), where the constraints and affordances are quite different from those in traditional settings. Such environments pose entirely different challenges and opportunities. In principle, assessments integrated with instruction can play a more salient role given concerns about attrition rates and academic achievement in online courses. Presumably, students in such courses have uninterrupted internet access so that they can access on-demand assessment at any time. Assessments that provide rich feedback and support for instruction would be especially valuable in such settings. There is little evidence, however, that assessment providers have risen to the occasion and developed portfolios of diverse assessment instruments that begin to meet these needs. Many e-learning environments offer chat rooms, threaded discussion capabilities and the like. Consequently, students can participate in multiple conversations, review archival transcripts and communicate asynchronously with teachers. These settings offer opportunities for collecting and analyzing data on the nature and patterns of interactions among students and teachers. Again, there is little evidence that such assessments are being undertaken.

It is also necessary to expand the notion of the virtual environment to include the cultural milieu of today's students as well as the work world of their parents and grandparents. Technology plays an important role in providing a seemingly endless stream of new tools (toys) to which individuals become accustomed in their daily lives at home and in the workplace. These range from cell phones and palm pilots to computers (and the sophisticated software applications that have been developed for them). Bennett (2002) discusses at length how the continuing infusion of new technology into the workplace, and the increasing importance of technology-related skills, places enormous pressure on schools both to incorporate technology into the curriculum and to enhance its role as a tool for instruction and assessment.

Another, equally fundamental, question is how children who grow up in an environment filled with continuous streams of audio-visual stimulation, and who may spend considerable time in a variety of virtual worlds, understand and represent reality, develop patterns of cognition and learning preferences. Some of these issues have been investigated by Turkle (1984, 1995). Increasingly, this will pose a challenge to both instruction and assessment. Teachers and educational software developers will have to take note of these (r)evolutionary trends if they are to be effective and credible in the classroom.

6. TECHNOLOGY AND PURPOSE

Although purpose is not directly impacted by technology, context does influence the relative importance of the different kinds of purpose – choose, learn or qualify. While all three are essential and will continue to evolve under the influence of technology and other forces, political and economic considerations will usually determine which one is paramount at a particular time and place. For a particular purpose, these same considerations will govern the direction and pace of technological change. As we have already argued, advances in the technology (and the science) of assessment make possible new trade-offs between validity and efficiency with respect to that purpose. However, the decisions actually taken may well leave the balance point unchanged, favour improvements in efficiency over those in validity, or vice versa.

This assertion is illustrated by the current focus in the U.S. on accountability testing, which has already been alluded to. Concern with the productivity of public education has led to a rare bi-partisan agreement on the desirability of rigorous, standards-based testing in several subjects at the end of each school year. As a result, hundreds of millions of dollars will be allocated to the states for the development and administration of these examinations and for the National Assessment of Educational Progress that is to be used (in part) to monitor the states' accountability efforts. Moreover, it appears that, in the short run at least, concerns with efficiency will take precedence over improvements in validity.

Recent surveys suggest that the U.S. public is generally in agreement with their elected officials. On the other hand, many educators and educational researchers have expressed grave concern about the trend to increased testing. They assert that increased reliance on high stakes external assessments will only stifle innovation and lead to lowered productivity. In other words, they are questioning the systemic validity (Frederiksen, & Collins, 1989) of such tests. Moreover, they argue that the effectiveness of high stakes tests as a tool for reform is greatly exaggerated and that these tests are favoured by those in government and business only because the required investments are relatively small in comparison to those required for improvements to physical plant or for meaningful changes in curriculum and instruction. See for example Amrein and Berliner (2002), Kohn (2000) and Mehrens (1998). Among those who support a greater role for technology in education, there is concern that the current environment presents many impediments to rapid progress, at least in the near term (Solomon & Schrum, 2002). In addition to the emphasis on high stakes testing, they cite economic, philosophical and empirical factors.

At bottom, there are strongly opposing views about the kinds of tests and types of technology that are most needed and, therefore, what is the most effective way to allocate scarce resources. How this conflict is resolved, and the choices that are made, will indeed determine which assessment technologies are developed, implemented and brought to scale.

7. CONCLUSIONS

The framework and corresponding analysis that have been presented strongly support the contention that technology has enormous potential to influence assessment. In brief, we have argued that the evolution of assessment practice results from the dynamic interplay between the demands on assessment imposed by a world that is shaped to some degree by technology and the range of possible futures for assessment made possible by the tools and systems that are the direct products of technology. Salomon and Almog (2000) have made an analogous argument with respect to the relationship between technology and educational psychology.

It is important to consider the negative impact technology can have on education generally and assessment practice, specifically. Policy makers see technology as a "quick fix" to the ills of education. Resource allocation is then skewed toward capital expenditures but, typically, without sufficient concomitant investments in technical support, teacher training and curriculum development. One consequence is inefficient usage (high opportunity costs), accompanied by demoralisation of the teaching force. Similarly, new technologies, such as item generation, can make multiple choice items – in comparison to performance assessments -- very attractive to test publishers and state education departments, with the result that assessment continues to be seen as the primary engine of change.

Too often, discussions of the promise of technology dwell on how it makes possible "new modes of assessment". For the most part, that is a misnomer. More typically, the role of technology is to facilitate the broader dissemination of assessment practices that were heretofore reserved for the fortunate few. Intelligent tutoring systems, for example, are often regarded as incorporating new modes of instruction and assessment. However, the notion that a learner ought to have the benefit of a teacher who has mastered all relevant knowledge, applies state-of-the-art pedagogical principles and adapts appropriately to the student's learning style, is not a new one. In point of fact, it is probably not very far from what Philip of Macedon had in mind when he hired Aristotle to tutor his son Alexander. Perhaps we should reserve the term "new modes" for such innovations as virtual reality simulations or assessments of an individual's ability to carry out information search on the web. Some other examples are found in Baker (2000).

It may be more productive to speculate on how the needs and interests of a technology-driven world may lead to new views of the nature and function of assessment. At the same time, further developments in cognitive science, educational psychology and psychometrics, along with new tool systems, undoubtedly will contribute to the evolution of assessment practice.

While educators and educational researchers are not ordinarily in a position to influence the macro trends that shape the environment, they are not without relevant skills and resources. They ought to cultivate a broader perspective on assessment and a deeper understanding of the interplay between context, purpose and assets. That understanding can help them to better predict trends in education generally and anticipate future demands on assessment, in particular. It may then be possible to formulate responses that substantially improve both validity and efficiency. Such responses may involve novel applications of existing technology or even the development of entirely new methods and the corresponding technologies for their implementation.

Those who are concerned about how assessment develops under the influence of technology and other forces must keep their eye on the potential of technology to democratise the practice of assessment. With validity and equity as lodestars, it is more likely that the power of technology can be harnessed in the service of humane educational ends.

REFERENCES

Amrein, A. L., & Berliner, D. C. (2002). High-stakes testing, uncertainty, and student learning. *Education Policy Analysis Archives, 10* (18).

Anderson, J. R., & Reiser, B. (1984). The LISP tutor. *Byte, 10*, 159-175.

Anderson, J. R., Boyle, C. F., & Yost, G. (1985). The geometry tutor. In A. Joshi (Ed.), *Proceedings of the Ninth International Joint Conference in Artificial Intelligence*. Los Altos, CA: Morgan Kaufman.

Bachman, L. F., & Palmer, A. S. (1996). *Language testing in practice*. Oxford: Oxford University Press.

Baker, E. L. (2000). *Understanding educational quality: Where validity meets technology*. Princeton, NJ: Educational Testing Service.

Bejar, I. I. (2002). Generative testing: From conception to implementation. In S .H. Irvine & P. C. Kyllonen (Eds.), *Item generation for test development* (pp. 199-217). Hillsdale, NJ: Lawrence Erlbaum Associates.

Bejar, I. I., & Braun, H. I. (1994). On the synergy between assessment and instruction: Early lessons from computer-based simulations. *Machine-Mediated Learning, 4*, 5-25.

Bejar, I. I., & Braun, H. I. (1999). *Architectural simulations – From research to implementation: Final report to the National Council of Architectural Registration Boards* (RM-99-2). Princeton, NJ: Educational Testing Service.

Bennett, R. E. (1998). *Reinventing assessment*. Princeton, NJ: Educational Testing Service.

Bennett, R. E. (2001). How the internet will help large-scale assessment reinvent itself. *Education Policy Analysis Archives, 9,* (5).

Bennett, R. E. (2002). Inexorable and inevitable: The continuing story of technology and assessment. (RM-02-03) Princeton, NJ: Educational Testing Service.

Braun, H. I. (2000). *Reflections on the future of assessment*. Invited paper presented at the 1st Conference of the EARLI/SIG assessment, Maastricht, The Netherlands.

Bunderson, C. V., Inouye, D. K., & Olsen, J. B. (1989). The four generations of educational measurement. In R. L. Linn (Ed.), *Educational Measurement* (3rd ed.) (pp. 367-407). New York: ACE/Macmillan.

Burstein, J. C., Wolff, S., & Lu, C. (1999). Using lexical semantic techniques to classify free responses. In N. Ide & J. Veronis (Eds.), *The depth and breadth of semantic lexicons*. Dordrecht: Kluwer Academic Press.

Evans, P., & Wurster, T .S. (2000). *Blown to bits: How the new economics of information transforms strategy*. Boston: Harvard Business School Press.

Falmagne, J. C., Doignon, J.-P., Koppen, M., Villano, M., & Johannesen, L. (1990). Introduction to knowledge-based spaces: How to build, test, and search them. *Psychological Review, 97* (2), 201-224.

Frederiksen, J., & Collins, A. (1989). A systems approach to educational testing. *Educational Researcher, 18* (9), 27-32.

Kohn, A. (2000). *The case against standardized testing: Raising the scores, ruining the schools.* Portsmouth, NH: Heineman.

Lesgold, A., Lajoie, S., Bunzo, M., & Egan, G. (1992).SHERLOCK: A coached practice environment for an electronics troubleshooting job. In J. H. Larkin & R. W. Chabay (Eds.), *Computer-assisted instruction and intelligent tutoring systems: Shared goals and complementary approaches* (pp. 201-238). Hillsdale, NJ: Lawrence Erlbaum Associates.

Mehrens, W. A. (1998). Consequences of assessment: What is the evidence? *Education Policy Analysis Archives, 6* (13).

Mislevy, R. J., & Gitomer, D. H. (1996). The role of probability-based inference in an intelligent tutoring system. *User modeling and user-adapted interaction, 5*, 253-282.

Mislevy, R. J., Almond, R. G., & Steinberg, L.S. (1999). Bayes nets in educational assessment: Where the numbers come from. In K. B. Laskey & H. Prade (Eds.), *Proceedings of the Fifteenth Conference on Uncertainty in Artificial Intelligence* (pp. 437-446). San Francisco: Morgan Kaufman.

Mislevy, R. J., Steinberg, L. S., Breyer, F. J., Almond, R. G., & Johnson, L. (1999). A cognitive task analysis with implications for designing simulation-based performance assessment. *Computers in Human Behavior, 15*, 335-374.

Mislevy, R. J., Steinberg, L. S., & Almond, R. G. (2002). On the roles of task model variables in assessment design. In S. Irvine & P. Kyllonen (Eds.), *Item Generation for Test Development* (pp. 97-128). Hillsdale, NJ: Lawrence Erlbaum.

Mitchell, R. (1992). *Testing for Learning: How new approaches to evaluation can improve American schools.* New York: The Free Press.

Pellegrino, J. W., Chudowski, N., & Glaser, R. (Eds.). (2001). *Knowing what students know: The science and design of educational assessment.* Washington, DC: National Academy Press.

Salomon, G., & Almog, T. (2000). Educational psychology and technology: A matter of reciprocal relations. *Teachers College Record, 100* (1), 222-241.

Shute, V. J., Torreano, L. A., & Willis, R. E. (2000). DNA: Towards an automated knowledge elicitation and organization tool. In S. P. Lajoie (Ed.), *Computers as Cognitive Tools*, Vol. 2 (pp. 309 – 335). Hillsdale, NJ: Lawrence Erlbaum.

Shute, V. J., & Torreano, L. A. (2001). Evaluating an automated knowledge elicitation and organization tool. In T. Murray, S. Blessing, & S. Ainsworth (Eds.), *Authoring tools for advanced technology learning environments: Toward cost-effective adaptive, interactive and intelligent educational software.* New York: Kluwer.

Snow, R. E., & Mandinach, E. B. (1999). Integrating assessment and instruction for classrooms and courses: *Programs and prospects for research.* Princeton, NJ: Educational Testing Service.

Solomon, G. & Schrum, L. (2002, May 29). Web-based learning. *Education Week.*

Swanson, L., & Stocking, M. L. (1993). A model and heuristic for solving very large item selection problems. *Applied Psychological Measurement, 17*, 151-166.

Turkle, S. (1984). *The second self: Computers and the human spirit.* New York: Simon and Schuster.

Turkle, S. (1995). *Life on the screen: Identity in the age of the internet.* New York: Simon and Schuster.

Wenger, E. (1987). *Artificial intelligence and tutoring systems.* Los Altos, CA: Morgan Kaufman.

Wolfe, M. B. W., Schreiner, M. E., Rehder, B., Laham, D., Foltz, P. W., & Landauer, T. K. (1998). Learning from text: Matching readers and texts by latent semantic analysis. *Discourse Processes, 25*, 309-336.

Index

Abbot, R. D. 71, 81
Abson, D. 74, 78
Acquah, S. 75, 84
Ahuna-Ka`ai, J. 146, 169
Ainsworth, S. 287
Alexander, R. A. 253, 265
Algozzine, B. 72, 81
Almog, T. 285, 287
Almond, R. G. 274, 276, 277, 287
Amdur, L. 24, 34
Amiran, M. R. 154, 166
Amrein, A. L. 284, 286
Anderson, A. 78
Anderson, J. R. 18, 33, 279, 286
Angoff, W. H. 249, 264
Anson, C. M. 144, 161
Anthony, R. 145, 161
Arnold, L. 74, 78
Arter, J. 148, 161
Arter, J. A. 24, 33, 144, 161, 165
Aschbacher, P. 155, 165
Aschbacher, P. R. 144, 155, 162, 165
Ashworth, P. 215, 221
Askham, P. 48, 52
Atkins, M. 15, 33

B.C. Ministry of Education 156, 162
Bachman, L. F. 274, 286
Bailey, J. J. 253, 265
Bain, J. 44, 47, 54

Baker, B. 75, 84
Baker, E. 4, 12, 38, 53, 60, 81, 129, 140
Baker, E. L. 148, 164, 285, 286
Bakewell, C. 74, 83
Bakker, J. 2, 12, 246
Bakshi, T. S. 93, 115
Bangert-Drowns, R. L. 66, 78
Bannister, P. 215, 221
Barnea, N. 103, 115
Barnett, J. E. 60, 64, 78, 85
Baron, J. B. 148, 162, 166
Barrows, H. S. 121, 139
Bartel, D. 168
Barth, R. 158, 162
Bass, K. M. 113, 116
Battistich, V. 228, 245
Baumert, J. 227, 245
Baxter, G. P. 90, 115, 117, 122, 140
Beatty, R. W. 57, 78
Bedeian, A. G. 57, 80
Bejar, I. I. 139, 275, 276, 286
Belanoff, P. 146, 149, 150, 161, 164, 166
Bell, D. 14, 33
Ben-Chaim, D. 61, 87, 213, 223
Ben-Elyahu, S. 101, 115
Benett, Y. 138, 139
Bennett, R. E. 268, 276, 283, 286
Benoit, J. 147, 162
Ben-Peretz, M. 100, 115
Bereiter, C. 113, 115
Bergee, M. J. 75, 78
Berliner, D. C. 284, 286
Bernadin, H. J. 57, 78
Berryman, L. 147, 149, 162
Bettenhausen, K. L. 57, 80
Biggs, J. 13, 15, 20, 21, 33, 44, 48, 52, 137, 139, 144, 162
Billett, S. 18, 33
Bintz, W. 155, 162
Birenbaum, M. 1, 2, 4, 8, 11, 13, 22, 24, 33, 34, 36, 37, 45, 48, 52, 61, 78, 90, 115, 120, 139, 140, 209, 218, 219, 221
Birenbaum,. M. 25
Birkeland, T. S. 73, 78
Biskin, G. H. 249, 265
Bixby, J. 22, 36, 45, 54, 149
Bixby, J.. 168
Black, L. 162, 163, 168

289

Black, P. 23, 24, 25, 29, 32, 34, 48, 52, 56, 78, 89, 92, 112, 113, 115, 152, 153, 162
Blank, L. L. 69, 84
Blatchford, P. 60, 78
Blessing, S. 287
Block, K. C. C. 168
Blum, R. 165
Blumenfeld, P. C. 91, 113, 115, 116
Blumer, H. 17, 34
Blythe, T. 23, 35, 156, 157, 162
Boekaerts, M. 6, 11, 35, 36, 225, 226, 245
Boersma, G. 63, 78
Boes, W. 196, 197, 221
Boix-Mansilla, V. 145, 164
Bond, L. 138, 139
Bonk, C. J. 21, 34
Boud, D. 55, 56, 57, 58, 60, 61, 79, 80, 139, 225, 245
Bouton, K. 70, 79
Bowden, J. 119, 139
Boyle, C. F. 279, 286
Boyston, J. A. 34
Bracken, B. A. 95, 116
Bradley, E. W. 61, 81
Brady, M. P. 72, 81
Brandt, R. S. 35
Braun, H. I. 10, 267, 269, 276, 286
Bray, G. B. 149, 165
Breyer, F. J. 277, 287
Broad, R. L. 147, 162
Broadfoot, P. 62, 86 159, 162
Brock, M. N. 75, 79
Brown, I. 75, 80
Brown, J. S. 17, 34
Brown, R. L. 144, 161
Brown, S. 9, 12, 47, 54, 55, 56, 79, 120, 129, 139, 140, 190, 205, 223
Bruner, J. S. 16, 21, 22, 35
Bucat, R. 91, 115
Bunderson, C. V. 268, 286
Bunzo, M. 279, 287
Burke, R. J. 76, 79
Burnett, W. 74, 79
Burstein, J. C. 276, 278, 286
Bush, E. S. 66, 80
Busick, K. 156, 162
Butler, D. L. 66, 79
Butler, R. 153, 162

Butterworth, R. W. 153, 162
Byard, V. 67, 79
Byrd, D. R. 70, 79
Byrne, B. 236, 245

Caine, G. 152, 162
Caine, R. N. 152, 162
Calenger, B. J. 116
Calfee, R. 125, 139, 162, 165
Calfee, R. C. 148, 162
Califano, L. Z. 70, 79
Calkins, V. 74, 78
Callahan, C. M. 253, 265
Callahan, S. F. 148, 163
Cameron, C. 142, 156, 163, 164, 166
Cannella, A. A. 57, 80
Carline, J. D. 69, 84
Carney, J. M. 145, 163
Carr, W. 152, 163
Carter, M. A. 144, 168
Cascallar, A. S. 10, 247, 263, 264
Cascallar, E. C. 1, 10, 247
Catterall, M. 69, 72, 79
Cavaye, G. 74, 79
Chabay, R. W. 287
Challis, M. 198, 221
Chambers, E. 209, 221
Champagne, A. B. 116
Chan, Yat Hung 147, 161, 163
Chaudron, C. 73, 79
Cheung, D. 91, 115
Chudowski, N. 280, 287
Churchill, G. A. 131, 139
Cicchetti, D. V. 57, 79
Cizek, G. J. 11, 247, 248, 264, 265, 266
Cobb, P. 17, 18, 34
Cohen, A. S. 251, 264
Cohen, E. G. 67, 79
Cohen, R. 57, 79
Cole, D. A. 66, 79
Collins, A. 17, 34, 38, 43, 53, 144, 164, 284, 287
Collins, C. 2, 12, 246
Collins, M. L. 64, 87
Collis, J. M. 246
Comfort, K. 90, 116
Como, L. 50, 52
Condon, W. 147, 163
Connell, J. P. 48, 53, 228, 245

Conway, R. 74, 80
Corno, L. 14, 36
Costa, A. 144, 149, 157, 163
Costa, A. L. 92, 115
Coulson, R. L. 119, 139
Cover, B. T. L. 70, 80
Covington, M. 150, 163
Cronbach, L. J. 38, 52
Crooks, T. 49, 52, 152, 163
Crooks, T. J. 56, 80, 251, 264
Culham, R. 148, 161
Cunningham, J. W. 16, 17, 34
Czerniac, C. 91, 117

Daiker, D. A. 162, 163, 168
Dambrot, F. H. 253, 265
Danielson, C. 157, 163
Darling-Hammond, L. 94, 115, 149, 157, 163
Davies, A. 6, 7, 141, 142, 146, 149, 163, 168
Davies, M. 164
Davies, P. 69, 80
Davis, J. K. 62, 80
Davis, M. H. 196, 222
De Corte, E. 119, 139
De Haan, D. H. 125
De Haan, D. M. 139
DeCharms, R. 150, 163
Deci 246
Deci, E. L. 48, 52, 53, 150, 163, 228, 229, 245, 246
Deibert, E. 148, 165
Delandshere, G. 33, 34
Dennis, I. 60, 83
Derry, S. J. 17, 34
Des Marchais, J. E. 121, 139
Desai, L. A. 144, 168
DeVoge, J. G. 147, 163
Dewey, J. 17, 34, 142, 152, 163
Dickson, M. 161, 164, 166
Dierick, S. 5, 33, 34, 37, 38, 39, 51, 52
Dillon, J. T. 92, 115
Dochy, F. 1, 2, 4, 5, 8, 11, 33, 34, 36, 37, 38, 39, 43, 45, 46, 47, 48, 49, 51, 52, 55, 61, 69, 73, 78, 80, 85, 90, 115, 120, 139, 140, 171, 172, 200, 201, 202, 221, 223
Doignon, J.-P. 279, 286

Dolmans, D. 125, 127, 139
Dori, Y. J. 6, 7, 89, 90, 92, 93, 95, 103, 106, 107, 109, 112, 113, 115, 116, 117, 118
Douglas, G. 91, 115
Dove, P. 55, 79
Downing, T. 75, 80
Drew, S. 204, 205, 221
Duffy, M. 154, 163
Duffy, T. M. 120, 121, 140
Duguid, P. 17, 34
Dumais, B. 121, 139
Dunbar, S. B. 4, 12, 32, 34, 38, 53, 129, 140
Dunne, E. 81
Durth, K. A. 145, 164
Duschl, R. A. 114, 116
Dweck, C. S. 56, 66, 80
Dwyer, W. O. 95, 116
Dyason, M. D. 57, 82
Dylan, W. 48, 52

Earl, L. 146, 164
Ebel, R. L. 248, 249, 263, 264
Eberhart, G. 74, 78
Eckes, M. 144, 165
Edelstein, R. A. 202, 203, 218, 221
Egan, G. 279, 287
Egharari, H. 229, 245
Ehly, S. W. 57, 66, 84, 86
Eisenhart, M. 30, 34
Elbow, P. 144, 146, 149, 150, 152, 164
El-Koumy, A. S. A. 62, 80
Elliot, A. 68, 86
Ellis, J. 64, 84
Engel, C. E. 121, 139
English, F. W. 125, 139
Entwistle, A. 15, 34, 191, 192, 209, 221
Entwistle, N. 11, 12, 53, 191, 192, 195, 218, 221, 222
Entwistle, N. J. 8, 11, 15, 34, 44, 47, 49, 52, 53, 171, 191, 192, 209, 221
Eresh, J. T. 60, 82, 144, 146, 148, 149, 165, 166
Ernest, P. 18, 34
Evans, P. 278, 286

Faggen, J. 250, 251, 264
Fagot, R. F. 233, 245

Falchikov, N. 38, 48, 53, 54, 55, 57, 60, 61, 67, 68, 69, 73, 74, 76, 79, 80, 129, 140, 225, 226, 233, 245
Falmagne, J. C. 279, 286
Farh, J. 57, 67, 80
Farivar, S. 66, 87
Fedor, D. B. 57, 80
Feldman, R. A. 209, 218, 219, 221
Feletti, G. 139
Feltovich, P. J. 119, 139
Fennimore, T. 2, 12, 246
Fenstermacher, G. D. 20, 34
Fernandes, M. 64, 80
Feuerstein, R. 17, 34
Fine, C. 2, 12, 246
Fitzgerald, J. 16, 17, 34
Fitzpatrick, A. R. 253, 264
Flanagan, J. C. 248, 249, 264
Flint, N. 221
Flood, J. 149, 164
Flores, G. 90, 116
Foltz, P. W. 278, 288
Fontana, D. 64, 80
Foot, H. C. 78
Ford, K. 143, 164
Fosnot, C. T. 18, 20, 34
Franklin, C. A. 75, 81
Franklyn-Stokes, A. 215, 216, 217, 221
Fraser, B. 115, 116
Fraser, B. J. 116
Fraser, B. J.. 118
Fredericksen, J. R. 38, 164
Frederiksen, J. 144, 284, 287
Frederiksen, J. R. 53
Frederiksen, R. J. 139
Fredricks, J. F. 113, 116
Freedman, S. W. 71, 81, 148, 162
Freeman, M. 69, 81
Freire, P. 15, 20, 34
Friedman Ben-David, M. 196, 199, 222
Fritz, C. A. 144, 147, 148, 156, 157, 158, 164
Fry, S. A. 68, 69, 81
Frye, A. W. 61, 81
Fu, D. 157, 164
Fuchs, D. A. 153, 164
Fuchs, L. S. 153, 164
Fullan, M. 157, 164
Furnham, A. 77, 81

Gagne, R. M. 264
Gaier, E. L. 61, 81
Gaillet, L. I. 55, 81
Gale, J. 36
Gallagher, J. J. 90, 118
Gallard, A. J. 100, 116
Gammon, L. 74, 78
Gao, X. 122, 140
Gardner, H. 17, 22, 35, 36, 45, 54, 141, 145, 149, 164, 168
Gearhart, M. 60, 81, 148, 150, 155, 164, 165, 166
Gere, A. R. 71, 81
Gergen, K. J. 17, 35
Gibbs, G. 44, 53, 129, 139
Gielen, S. 5, 37
Gifford, B. R. 117
Gijbels, D. 172, 221
Gillespie, C. 143, 144, 145, 157, 164
Gillespie, R. 143, 164
Gipps, C. 22, 35, 144, 152, 164
Gipps, P. 50, 51, 53
Gitmore, D. H. 114, 116
Gitomer, D. A. 146, 148, 149, 166
Gitomer, D. H. 60, 82, 279, 287
Glaser, R. 124, 139, 248, 264, 280, 287
Glasner, A. 139
Glatthorn, A. A. 14, 35
Glenn, J. 22, 36, 45, 54, 149, 168
Globerson, T. 67, 85
Goicoechea, J. 18, 35
Goldfinch, J. 55, 69, 74, 80, 81, 225, 226, 233, 245
Goldman, S. R. 90, 115
Goleman, D. 150, 164
Goodlad, S. 86
Gorsuch, R. L. 212, 223
Graham, S. 70, 83, 100, 116
Graner, M. H. 73, 81
Greene, D. 150, 166
Greeno, J. G. 18, 33
Gregory, K. 142, 146, 157, 163, 164
Griffee, D. T. 61, 81
Griffin, C. 14, 35
Griffiths, S. 79, 82, 84, 86, 87
Grobler, S. 223
Grosso, L. 253, 265
Gruppen, L. 61, 87

Index

Guba, E. 143, 166
Guba, E. G. 16, 35, 91, 116
Gunnar, M. R. 245
Gunstone, R. F. 63, 85
Gutvirtz, Y. 221

Haaga, D. A. F. 69, 81
Haaga, D. A. F. 69
Haertel, E. H. 38, 53
Hambleton, R. K. 10, 11, 249, 253, 263, 264
Hamp-Lyons, L. 147, 163
Hansen, J. 149, 165
Harari, H. 107, 116
Harasim, L. 21, 35
Harden, R. M. 196, 222
Hargreaves, A. 164
Harlen, W. 113, 116
Harste, J. 155, 162
Hartley, F. E. 166
Harvey, A. L. 254, 265
Hattie, J. 91, 115, 152, 153, 165
Hawkins, S. 233, 246
Hayes Maeshiro, M. 146, 169
Helmore, P. 69, 83
Henderson, H. W. 74, 83
Hendrickson, J. M. 72, 81
Henry, S. E. 66, 81
Herman, J. 2, 12, 60, 81, 148, 149, 164, 165, 246
Herman, J. L. 144, 148, 155, 165, 166
Herried, C. F. 92, 93, 116
Herrmann, A. 35
Herscovitz, O. 107, 109, 116
Hertz, N. R. 253, 265
Hertz-Lazarowitz, R. 94, 116
Heun, L. R. 74, 81
Heywood, J. 60, 81
Hidi, S. 227, 245
Hixon, J. E. 60, 64, 78, 85
Hoffman, L. 227, 245
Hofstein, A. 93, 116
Holford, J. 14, 35
Holley, C. A. B. 70, 81
Holst, P. M. 56, 85
Hoover, H. D. 32, 34, 149, 165
Horsch, E. 45, 53
Hounsell, D. 8, 11, 12, 53, 55, 81, 218, 221, 222

Hounsell, D. J. 44, 53
Houston, K. 79, 82, 84, 86, 87
Howard, K. 144, 146, 149, 150, 157, 165
Howe, C. J. 78
Howie, P. W. 196, 222
Huffman, D. 90, 116
Hughes, I. E. 69, 73, 81, 82
Hunter, D. 75, 82
Hurtz, G. M. 253, 265

Ide, N. 286
Impara, J. C. 10, 12, 249, 263, 265
Ingenkamp, K. 223
Ingham, D. 64, 87
Inouye, D. K. 268, 286
Irvine, S. 287
Irvine, S. H. 286
Irwin, P. M. 249, 265

Jacobowitz, R. 90, 118
Jacobs, G. 73, 82
Jaeger, R. M. 248, 252, 263, 264, 265
Jager, R. S. 223
Jailall, J. 14, 35
Janssens, S. 171
Jarvis, P. 14, 35
Jenkins, E. W. 92, 117
Johannesen, L. 279, 286
Johnson, L. 277, 287
Johnson, T. 145, 161
Joines, S. M. B. 57, 82
Jones, B. F. 2, 12, 246
Jones, J. 154, 163
Jones, L. R. 10, 12, 263, 265
Jöreskog, K. G. 233, 245
Joshi, A. 286
Joslin, G. 153, 157, 165
Julius, T. M. 144, 149, 151, 154, 165

Kaberman, Z. 103, 115
Kallick, B. 144, 149, 157, 163
Kamii, C. 153, 165
Kane, M. 38, 53, 253, 263, 265
Kane, M. T. 251, 264
Karegianes, M. L. 69, 70, 82
Kaye, A. 35
Kaye, M. M. 57, 82
Keiny, S. 91, 116, 118
Keller, J. M. 48, 53

Kember, D. 74, 80
Kemmis, S. 152, 163
Ker, J. 196, 222
Kimura, A. 146, 169
Kinchin, G. D. 145, 165
King, K. S. 21, 34
Klaus, D. J. 248, 264
Klein, S. 22, 32, 35, 60, 82, 148, 165
Kleinasser, A. 45, 53
Klenowski, V. 145, 157, 165
Knight, P. 56, 79, 120, 139
Knight, P. K. 33
Kniveton, B. H. 196, 203, 204, 222
Kohn, A. 152, 153, 165, 284, 287
Konet, R. J. 145, 165
Koppen, M. 279, 286
Koretz, D. 22, 23, 32, 35, 60, 82, 148, 165
Koretz, D. M. 32, 34
Korman, M. 69, 82
Krajcik, J. 91, 113, 115, 116
Krapp, A. 227, 242, 245
Kuhs, T. 144, 165
Kulhavy, R. W. 66, 82
Kulieke, M. 2, 12, 244, 246
Kulik, C. C. 66, 78
Kulik, J. A. 66, 78
Kulski, M. M. 35
Kwun, S. K. 76, 85
Kyllonen, P. 287
Kyllonen, P. C. 286

Lagana, J. R. 70, 82
Laham, D. 278, 288
Lajoie, S. 279, 287
Lajoie, S. P. 287
Lamme, L. 157, 164
Landauer, T. K. 278, 288
Lapp, D. 149, 164
Large, B. J. 69, 82
Larkin, J. H. 287
Lasater, C. A. 75, 82
Latane, B. 233, 246
Lave, J. 17, 35
Law, B. 144, 165
Lawrence, M. J. 69, 82
Lawrenz, F. 90, 113, 116
Lazarowitz, R. 92, 93, 94, 115, 116, 117
Lazear, D. 145, 165

Lazenblatt, A. 79, 82, 84, 86, 87
Le Doux, J. 150, 165
Lee, B. 63, 82
Leeming, F. C. 95, 116
Leinhardt, G. 125, 140
Lejk, M. 69, 82
LeMahieu, P. 60, 82, 141, 144, 146, 148, 149, 150, 157, 164, 165, 166
LeMahieu, P. G. 144
Lennon, S. 75, 77, 82
Lens, W. 45, 54
Leone, D. R. 229, 245
Lepper, M. R. 150, 166
Lesgold, A. 279, 287
Lewin, K. 142, 166
Lewis, S. 158, 166
Lewy, A. 89, 117
Lienert, G. A. 234, 246
Lincoln, Y. 143, 166
Lincoln, Y. S. 91, 116
Lindquist, E. F. 264
Linn, R. L. 4, 12, 32, 35, 38, 52, 53, 129, 140, 286
Liston, D. P. 157, 169
Litwack, M. 74, 82
Lloyd, J. 63, 83
Lomax, R. G. 222
Longhurst, N. 60, 83
Lovitts, B. E. 116
Lu, C. 276, 286
Luca, J. 21, 35
Lushene, R. E. 212, 223
Lynch, J. H. 228, 246
Lynn, R. L. 265
Lyons, N. 167

MacArthur, C. A. 70, 71, 83, 86
Macdonald, C. 146
MacDonald, C. 166
MacGowan, A. 60, 86
MacKenzie, L. 69, 83
MacLellan, E. 56, 83
Macnaughton, D. 142, 163
MacNaughton, D. 166
MacRury, K. 223
Maehr, M. 150, 166
Mager, R. F. 150, 166
Magin, D. 69, 83
Magin, D. J. 67, 69, 83

Index

Magnieri, J. N. 168
Mahoney, M. J. 150, 166
Mandinach, E. B. 279, 287
Manning, M. 144, 166
MAPP 156, 166
Marcoulides, G. A. 67, 69, 83
Marienau, C. 62, 83
Marlin, J. W. Jr. 222
Marton, F. 11, 12, 44, 47, 53, 119, 139, 191, 192, 193, 194, 218, 221, 222
Marx, R. W. 91, 113, 115, 116
Mason, G. 64, 87
Mason, R. 35
Mathews, B. P. 69, 74, 83
Maurer, T. J. 253, 265
Mayer, R. E. 17, 35
Mc Caffrey, D. 60, 82
Mc Dowell, L. 37, 52
McAuley, R. G. 74, 83
McCaffrey, D. 22, 32, 35, 148, 165
McCann, J. 150, 166
McClung, M. S. 125, 140
McCulloch, M. 55, 81
McCune, V. 191, 221
McCurdy, B. L. 72, 83
McDonald, B. 64, 83
McDonald, J. 91, 117
McDowell, L. 9 12, 47, 50, 53, 54, 129, 140, 190, 205, 208, 209, 222, 223
McLaughlin, M. 145, 157, 166
McLoughlin, C. 21, 35
Meece, J. L. 167
Mehrens, W. A. 248, 265, 284, 287
Melican, G. M. 249, 265
Merry, S. 69, 84, 198, 199, 222
Messick, S. 4, 12, 38, 43, 53, 246
Meyer, C. 150, 161, 166
Meyer, C. A. 71, 84
Meyer, D. K. 197, 198, 222
Michael, W. B. 153, 162
Mickelson, N. 145, 161
Midgley, D. F. 58, 83
Miller, C. M. 47, 53
Miller, M. 63, 83
Millman, J. 144, 157, 163, 166, 167, 249, 265
Mills, C. N. 249, 263, 264, 265
Mills, L. 63, 83
Mills-Court, K. 154, 166

Minnaert, A. 225
Mires, G. J. 199, 200, 222
Mislevy, R. J. 139, 274, 276, 277, 279, 287
Mitchell, K. J. 10, 12, 263, 265
Mitchell, R. 92, 113, 117, 276, 287
Mitchell, V. W. 74, 83
Miyasaka 157, 167
Mockford, C. D. 69, 83
Moerkerke, G. 33, 34, 43, 47, 48, 52
Morgan, M. 66, 78
Moss, P. A. 33, 35, 129, 140
Murray, T. 287

Nevo, D. 45, 53, 89, 90, 92, 95, 99, 112, 117
Newcomb, F. E. 166
Newman, R. S. 66, 84
Newstead, S. 60, 83
Newstead, S. A. 215
Newstead, S. E. 60, 83, 216, 217, 221
Ney, J. W. 72, 84
Ninness, H. A. C. 64, 84
Ninness, S. K. 64, 84
Nisan, M. 153, 162
Norcini, J. J. 253, 265
Norton, L. S. 60, 83
Novak, J. 148, 150, 164, 166
Novak, J. R. 148, 166
Novick, M. R. 249, 264
NRC - National Research Council 90, 117
Nuy, H. J. P. 121, 140

O'Connor, M. C. 117
O'Boyle, M. 64, 87
O'Donnell, A. M. 56, 70, 84, 85
Olsen, J. B. 268, 286
Olson, D. R. 16, 21, 22, 35
Olson, V. L. B. 71, 84
Orpen, C. 69, 84
Orr, A. 75, 84
Orsmond, P. 69, 84, 198, 199, 222

Packer, M. J. 18, 35
Palmer, A. S. 274, 286
Pankratz, R. 158, 167
Paris, A. H. 66, 84
Paris, S. G. 66, 84

Parker, J. 90, 118
Parlett, M. 47, 53
Pascarella, E. T. 69, 82
Patelis, T. 251, 266
Patrick, B. C. 229, 245
Paulson, E. L. 71
Paulson, E.L. 84
Paulson, F. L. 166
Paulson, F.L. 150, 161
Paulson, P. R. 71, 84, 150, 161, 166
Pellegrini, A. D. 223
Pellegrino, J. 280, 287
Pellegrino, J. W. 10, 12, 263, 265
Pelletier, L. G. 228, 245
Perfumo, P. 162, 165
Perkins, D. N. 23, 35
Petrosky, A. R. 33, 34
Petty, M. 58, 83
Pflaum, S. W. 70, 82
Phillips, D. C. 18, 35
Philp, J. R. 61, 81
Pierson, H. 70, 84
Pine, J. 90, 115
Pintrich, P. R. 14, 35, 36
Pippard, M. J. 196, 222
Pizzini, E. L. 92, 117
Plake, B. S. 10, 12, 249, 263, 264, 265
Poikela, E. 121, 140
Poikela, S. 121, 140
Politano, C. 142, 144, 163, 166
Pond, K. 67, 69, 84, 93, 117
Popam, W. J. 32, 35
Posch, P. 93, 117
Potter, E. F. 144, 166
Powney, J. 131, 140
Prawat, R. S. 17, 36
Preece, A. 145, 161
Preece, P. E. 199, 222
Prosser, M. 44, 47, 53, 54, 191, 195, 223
Purkey, W. 150, 166

Quesada, A. 145, 167

Raack, L. 2, 12, 246
Raatz, U. 234, 246
Rada, R. 69, 75, 84, 85
Ramaprasad, A. 23, 36
Ramsden, P. 15, 36, 44, 47, 53, 129, 140, 191, 192, 194, 195, 221, 222

Ramsey, P. 69, 75, 84, 85
Ramsey, P. G. 69, 75, 84
Rand, D. C. 62, 80
Randi, J. 14, 36
Raphael, T. E. 70, 84
Raymond, M. R. 261, 265
Reckase, M. 158, 167
Reder, L. M. 18, 33
Redfield, D. 158, 167
Reeder, E. 72, 85
Reese, B. F. 145, 167
Regehr, G. 61, 87
Rehder, B. 278, 288
Reich, R. 74, 84
Reid, H. M. 202, 221
Reid, J. B. 261, 265
Reigeluth, C. M. 53
Reiling, K. 69, 84
Reiser, B. 279, 286
Renninger, K. A. 227, 245
Resnick, D. 147, 148, 158, 167
Resnick, D. P. 92, 117
Resnick, L. 147, 148, 158, 167
Resnick, L. B. 92, 117
Richard, A. 161, 167
Richards, B. F. 61, 81
Richards, M. 156, 167
Richardson, J. T. E. 222
Richer, D. L. 73, 84
Richter, S. E. 144, 167
Rickinson, B. 213, 214, 222
Rijlaarsdam, G. 71, 84, 85
Riley, S. M. 67, 85
Roberts, W. H. 73, 85
Robin, F. 251, 266
Rocklin, T. R. 56, 85
Rogers, C. R. 59, 85
Rohrkemper, M. 50, 52
Rosenfeld, S. 93, 116
Ross, J. A. 72, 85
Ross, S. 61, 85
Rowntree, D. 60, 85
Rudd, T. J. 63, 85
Ruiz-Primo, M. A. 90, 117
Rushton, C. 69, 75, 85
Russ, M. 75, 82
Russell, D. 147, 149, 162
Ryan 157, 167

Index

Ryan, R. M. 48, 53, 150, 163, 228, 245, 246

Saavedra, R. 76, 85
Sadler, D. A. 50
Sadler, D. R. 54
Sadler, R. 23, 36, 144, 152, 153, 167
Salend, S. J. 72, 85
Säljö, R. 47, 53, 191, 192, 193, 194, 222
Salomon, G. 67, 85, 285, 287
Sambell, K. 9, 12, 47, 48, 54, 129, 130, 131, 140, 190, 205, 206, 207, 208, 209, 219, 222, 223
Sampson, J. 57, 79
Samway, K. D. 70, 85
Sanders, W. L. 58, 86
Santamaria, P. 150, 168
Saracho, N. O. 223
Sarason, I. G. 210, 212, 218, 223
Sarason, S. 112, 117
Sarig, G. 24, 36
Savery, J. R. 120, 121, 140
Saxe, L. 168
Scardamalia, M. 113, 115
Schaps, E. 228, 245
Schenk, S. M. 47, 54, 171, 217, 223
Schiefele, U. 227, 246
Schlechty, P. 157, 167
Schmitt, A. 9, 10, 12, 252, 253, 259, 261, 262, 265
Schmoker, M. 157, 167
Schoemaker, P. J. H. 124, 125, 140
Schon, D. A. 167
Schön, D. A. 26, 36, 142
Schonberger, L. C. 145, 167
Schoonen, R. 71, 85
Schotta, C. 64, 84
Schreiner, M. E. 278, 288
Schrum, L. 284, 287
Schunk, D. 54
Schunk, D. H. 14, 26, 36, 59, 85, 153, 167
Schwartz, S. S. 70, 83
Scouller, K. 44, 47, 54
Seagoe, M. V. 153, 167
Seewald, A. M. 125, 140
Segers, M. 1, 6, 8, 52, 55, 69, 73, 80, 85, 119, 122, 140, 172, 200, 201, 202, 221, 223

Seidel, S. 149, 168
Senge, P. 157, 167
Senge, P. M. 142, 152, 157, 167
Sfard, A. 18, 20, 36
Shapiro, E. S. 72, 83
Shavelson, R. 39, 54
Shavelson, R. J. 90, 115, 117, 122, 140
Shea, J. 253, 265
Shepard, L. 152, 153, 167
Shepard, L. A. 248, 265
Shepardson, D. P. 92, 117
Sherman, S. 64, 84
Shore, L. M. 76, 85
Shore, T. H. 76, 85
Shorrocks-Taylor, D. 92, 117
Shulman, L. 157, 167
Shulman, L. S. 107, 117
Shute, V. J. 275, 287
Simkin, M. G. 67, 69, 83
Simon, H. A. 18, 33
Sink, C. A. 64, 85
Sireci, S. G. 249, 251, 265, 266
Sivan, A. 74, 80
Sizer, T. 95, 117
Sizer, T. R. 142, 146, 168
Slater, T. F. 196, 223
Slattery, J. 60, 86
Sluijsmans, D. 52, 55, 80
Smeck, R. R. 52
Smith, A. 144, 168
Smith, B. 199, 222
Smith, E. F. 68, 86
Smith, J. K. 56, 87
Smith, M. 153, 167
Snow, R. E. 279, 287
Sobral, D. T. 62, 85
Solomon, D. 228, 245
Solomon, G. 284, 287
Solomon, J. 91, 93, 117
Soloway, E. 91, 113, 115, 116
Sommerich 82
Sommerich, C. M. 57, 82
Sommers, J. 162, 163, 168
Sörbom, D. 233, 245
Spandel, V. 24, 33, 144, 148, 161
Spielberger, C. D. 212, 223
Spiro, R. J. 119, 139
Spodek, B.. 223
Sroufe, L. A. 245

Stecher, B. 22, 32, 35, 60, 82, 148, 165
Stefani, L. A. J. 69, 72, 74, 76, 85
Steffe, L. P. 36
Steinberg, L. S. 274, 276, 277, 287
Sternberg 221
Sternberg, R. J. 17, 36
Stewart, G. 100, 116
Stiggins, R. 146, 150, 152, 161, 168
Stiller, J. 228, 246
Stobart, G. 152, 164
Stock, W. A. 66, 82
Stocking, M. L. 276, 287
Stoddard, B. 71, 86
Strachan, I. B. 74, 86
Stringfield, P. 77, 81
Struyf, E. 45, 54
Struyven, K. 171
Stubblefield, R. L. 69, 82
Stygall, G. 162, 163, 168
Sumida, A. 146, 169
Supovitz, J. A. 60, 86
Sutton, R. 152, 157, 168
Swanson, G. E. 166
Swanson, I. 68, 86
Swanson, L. 276, 287
Sweeney, A. E. 100, 117
Sylwester, R. 152, 168

Tait, H. 195, 221
Tal, R. T. 92, 93, 94, 95, 96, 99, 112, 113, 116, 117
Tamir, P. 89, 90, 118
Tanner, P. A. 150, 151, 155, 168
Tate, B. 142, 163, 166
Tatsuoka, K. K. 221
Taylor, I. 69, 86
Thomas, P. 44, 47, 54
Thomas, S. W. 154, 163
Thome, C. C. 152, 168
Thomson, K. 48, 54, 129, 140
Thorndike, R. L. 264
Thornton, G. C. 76, 85
Tierney, R. J. 144, 148, 149, 168
Tinzmann, M. B. 2, 12, 246
Tjosvold, D. 150, 168
Tobin, K. 100, 115, 116, 117
Tobin, K. G. 116, 118
Tolmie, A. K. 78
Topping, K. 172, 223

Topping, K. J. 6, 55, 57, 58, 66, 68, 69, 70, 84, 86
Torrance, N. 35
Torreano, L. A. 275, 287
Towler, L. 62, 86
Traub, R. E. 223
Treadwell, I, 223
Treagust, D. F. 90, 91, 114, 118
Trigwell, K. 44, 47, 53, 54, 191, 195, 223
Turkle, S. 283, 287
Turner, R. F. 75, 86
Tusin, L. F. 197, 198, 222
Tustad, S. 45, 53
Tutty, G. 70, 79
Tynjälä, P. 15, 36

Ul-Haq, R. 67, 84, 117
Usatine, R. 202, 221

Valdez, P. S. 144, 168
Vallerand, R. J. 228, 245
Van den Bossche, P. 172, 221
Van IJsendoorn, M. H. 172, 223
Van Rossum, E. J. 47, 54, 171, 217, 223
Vandenberghe, R. 45, 54
Veronis, J. 286
Viggiano, E. 100, 116
Vigiliano, M. 100, 116
Villano, M. 279, 286
Vogt, M. 145, 157, 166
Von Glasersfeld, E. 17, 36
Vosniadou, S. 18, 36
Vygotsky, L. S. 17, 36

Wade, L. K. 70, 86
Wade, W. 67, 84
Walker, P. 191, 221
Wallace, Jr., R. C. 146, 165
Walters, J. 149, 168
Wante, D. 196, 197, 221
Ward, M. 61, 87
Warden, D. A. 78
Wassef, A. 64, 87
Watson, M. 228, 245
Watts, J. 131, 140
Way, W. D. 254, 265
Weaver, M. E. 71, 87
Weaver, S. D. 145, 168
Webb, N. M. 66, 87

Index

Weeks, J. O. 70, 87
Weiser, I. 145, 169
Welch, W. 90, 116
Wellborn, J. G. 228, 245
Wenger, E. 17, 35, 279, 287
Wenrich, M. D. 69, 84
White, E. M. 147, 168
White, M. B. 70, 87
Whittaker, A. K. 148, 164
Whittaker, C. R. 72, 85
Wiggins, G. 92, 118, 168
Wilcox, S. 74, 86
Wiliam, D. 23, 24, 25, 29, 32, 34, 56, 78, 152, 153, 162
Wilkes, M. S. 202, 221
William, D. 9, 12, 263, 266
Williams, G. C. 228, 246
Williams, J. 74, 87
Williams, K. 233, 246
Willis, D. J. 144, 147, 151, 152, 156, 158, 168
Willis, R. E. 275, 287
Willoughby, L. 74, 78
Winne, P. H. 66, 79
Winters, L. 144, 165
Wittrock, M. C. 34
Wolf, D. 22, 36, 45, 54, 149, 168
Wolf, D. P. 166
Wolf, K. 157, 169
Wolf, S. 150
Wolfe, L. 56, 87
Wolfe, M. B. W. 278, 288
Wolfe, S. 164
Wolff, S. 276, 286
Wong-Kam, J. 146, 169
Worthen, B. R. 23, 36, 112, 118
Wright, L. 77, 87
Wu, M. 74, 80
Wurster, T. S. 278, 286
Wyvill, M. 69, 82

Yackel, E. 17, 18, 34
Yancey, K. B. 145, 154, 169
Yang, H. 147, 162
Yates, J. A. 75, 87
Yorke, M. 87
Yost, G. 279, 286
Young, C. A. 145, 152, 169

Zeichner, K. M. 157, 169
Zeidner, M. 35, 36, 209, 219, 223
Zhang 221
Zhang, S. 73, 82
Zieky, M. J. 249, 250, 251, 252, 263, 266
Zimmerman, B. J. 14, 26, 36
Zimmerman, B. K. 36
Zohar, A. 92, 118
Zoller, U. 92, 118, 212, 213, 223
Zoller, Z. 61, 87